Despeckle Filtering Algorithms and Software for Ultrasound Imaging

Despeckle Filtering Algorithms and Software for Ultrasound Imaging
Christos P. Loizou and Constantinos S. Pattichis
www.morganclaypool.com

ISBN: 9781598296204 paperback

ISBN: 9781598296211 ebook

DOI: 10.2200/S00116ED1V01Y200805ASE001

A Publication in the Morgan & Claypool Publishers series

SYNTHESIS LECTURES ON ALGORITHMS AND SOFTWARE IN ENGINEERING #1

Lecture #1

Series Editor: Andreas Spanias, Arizona State University

Series ISSN pending with U.S. Library of Congress

Despeckle Filtering Algorithms and Software for Ultrasound Imaging

Christos P. Loizou
Intercollege, Cyprus

Constantinos S. Pattichis
University of Cyprus

SYNTHESIS LECTURES ON ALGORITHMS AND SOFTWARE IN ENGINEERING #1

MORGAN & CLAYPOOL PUBLISHERS

ABSTRACT

It is well-known that speckle is a multiplicative noise that degrades image quality and the visual evaluation in ultrasound imaging. This necessitates the need for robust despeckling techniques for both routine clinical practice and teleconsultation. The goal for this book is to introduce the theoretical background (equations), the algorithmic steps, and the MATLAB™ code for the following group of despeckle filters: linear filtering, nonlinear filtering, anisotropic diffusion filtering and wavelet filtering. The book proposes a comparative evaluation framework of these despeckle filters based on texture analysis, image quality evaluation metrics, and visual evaluation by medical experts, in the assessment of cardiovascular ultrasound images recorded from the carotid artery. The results of our work presented in this book, suggest that the linear local statistics filter *DsFlsmv*, gave the best performance, followed by the nonlinear geometric filter *DsFgf4d*, and the linear homogeneous mask area filter *DsFlsminsc*. These filters improved the class separation between the asymptomatic and the symptomatic classes (of ultrasound images recorded from the carotid artery for the assessment of stroke) based on the statistics of the extracted texture features, gave only a marginal improvement in the classification success rate, and improved the visual assessment carried out by two medical experts. A despeckle filtering analysis and evaluation framework is proposed for selecting the most appropriate filter or filters for the images under investigation. These filters can be further developed and evaluated at a larger scale and in clinical practice in the automated image and video segmentation, texture analysis, and classification not only for medical ultrasound but for other modalities as well, such as synthetic aperture radar (SAR) images.

KEYWORDS

Speckle, despeckle, noise filtering, ultrasound, ultrasound imaging, cardiovascular imaging, SAR, texture, image quality, carotid artery

Dedication

To my wife, Phaedra; my son, Panayiotis; my daughter, Andrea; and my parents, Panayiotis and Eleni Loizou

Christos P. Loizou

To my family

Constantinos S. Pattichis

"Show thyself in all things an example of good works,
in teaching, in integrity and dignity;
let thy speech be sound and blameless,
so that anyone opposing may be put to shame,
having nothing bad to say for us.
Exhort slaves to obey their masters,
pleasing them in all things and not opposing them."

Titus 2: 7–9

Preface

Speckle is a multiplicative noise that degrades image quality and the visual evaluation in ultrasound and SAR imaging. This necessitates the need for robust despeckling techniques in a wide spectrum of the aforementioned imaging applications. Despeckle filtering applications has been a rapidly emerging research area in recent years. The goal for this book is to introduce the theoretical background (equations), the algorithmic steps, and the MATLAB™ code for the following group of despeckle filters: linear filtering, nonlinear filtering, anisotropic diffusion filtering, and wavelet filtering. The filters covered represent only a snapshot of the vast number of despeckle filters published in the literature. Moreover, selected representative applications of image despeckling covering a variety of ultrasound image processing tasks are presented. Most importantly, a despeckle filtering and evaluation protocol is proposed based on texture analysis, image quality evaluation metrics, and visual evaluation by experts. The source code of the algorithms presented in this book has been made available on the web, thus enabling researchers to more easily exploit the application of despeckle filtering in their problems under investigation.

The book is organized in six chapters. Chapter 1 presents a brief overview of ultrasound imaging, speckle noise, modeling, and filtering. In Chapter 2, the theoretical background (equations), the algorithmic steps, and the MATLAB™ code of selected despeckle filters are presented. Chapter 3 covers the material and recording of ultrasound images, and the evaluation methodology based on texture and statistical analysis, image quality evaluation metrics, and the experiments carried out for visual evaluation. Chapter 4 presents the applications of despeckle filtering techniques in ultrasound images of the carotid and cardiac ultrasound images. Chapter 5 discusses, compares, and evaluates the proposed despeckle filtering techniques where strong and weak points for each filtering technique are presented. Chapter 6 presents the summary and future directions, where a despeckling filtering protocol is also proposed. Finally, at the end of this book, an appendix provides details about the despeckle filtering MATLAB™ toolbox which can also be downloaded at http://www.medinfo.cs.ucy.ac.cy.

This book is intended for all those working in the field of image and video processing technologies, and more specifically in medical imaging and in ultrasound image and video preprocessing and analysis. It provides different levels of material to researchers, biomedical engineers, computing engineers, and medical imaging engineers interested in developing imaging systems with better quality images, limiting the corruption of speckle noise.

Acknowledgments

We thank all the members of our carotid ultrasound imaging team for the long discussions, advice, encouragement, and constructive criticism they provided us during the course of this research work. First of all, we express our sincere thanks to Emeritus Prof. Andrew Nicolaides of the Faculty of Medicine, Imperial College of Science, Technology and Medicine, London, UK, and founder of the Vascular Screening and Diagnostic Centre, Nicosia, Cyprus. Furthermore, we thank Dr. Marios Pantziaris, consultant neurologist at the Cyprus Institute of Neurology & Genetics, Nicosia, Cyprus, Dr. Theodosis Tyllis, consultant physician in the private sector in Cyprus, Associate Professor Efthyvoulos Kyriakou at the Frederick University, Lemesos, Cyprus, Dr. Christodoulos Christodoulou, research associate at the Nicosia, University of Cyprus, and Associate Professor Marios Pattichis, University of New Mexico, Las Cruces, NM. Last but not the least, we would like to thank Prof. Andreas Spanias, Arizona State University, Tempe, AZ, for his proposal and encouragement in writing this book and Joel Claypool and the rest of the staff at Morgan and Claypool Publishers for their understanding, patience, and support in materializing this project.

This work was partly funded through the projects Integrated System for the Support of the Diagnosis for the Risk of Stroke (IASIS, 2002–2005), and Integrated System for the Evaluation of Ultrasound Imaging of the Carotid Artery (TALOS, 2003–2005), funded by the Research Promotion Foundation of Cyprus. Furthermore, partial funding and support was also obtained from the Cardiovascular Disease Educational and Research Trust (CDER Trust) of UK and Cyprus.

We hope that this book will be a useful reference for all the readers in this important field of research and contribute to the development and implementation of innovative imaging and video systems enabling the provision of better quality images.

Contents

CHAPTER 1

Introduction to Ultrasound Imaging and Speckle Noise

According to an old Chinese proverb, "a picture is worth a thousand words." In the modern age, this concept is still true for computer vision and image processing tasks, where we aim to derive better systems and tools that give us different perspectives on the same image, thus allowing us to understand not only its content but also its meaning and significance. Image processing cannot compete with the human eye in terms of accuracy, but it can perform better on observational consistency and ability to carry out detailed mathematical operations. In the course of time, image-processing research has evolved from basic low-level pixel operations to high-level analysis that now includes sophisticated techniques for image interpretation and analysis. These new techniques are being developed to gain a better understanding of images based on the relationships between its components, context, history, and knowledge gained from a range of sources.

The purpose of this chapter is to give a brief overview of ultrasound imaging and present its basic principles and limitations. Furthermore, speckle noise is introduced as a major noise factor, which limits image resolution and hinters further image processing analysis in ultrasound images. We then introduce different despeckle filtering techniques that may be applied as a preprocessing step for denoising of ultrasound images. A few examples of despeckle filtering for real ultrasound images are given, and some of its limitations are discussed. Finally, at the end of this chapter, we present the statistics of speckle noise and its mathematical model.

1.1 A BRIEF REVIEW OF ULTRASOUND IMAGING

Medical imaging technology has experienced a dramatic change in the last 30 years. Previously, only X-ray radiographs were available, which showed the organs as shadows on a photographic film. With the advent of modern computers and digital imaging technology, new imaging modalities like computer tomography (CT or computer-assisted tomography), magnetic resonance imaging (MRI), positron emission tomography (PET), and ultrasound, which deliver cross-sectional images of a patient's anatomy and physiology, have been developed. Among the imaging techniques employed are X-ray angiography, X-ray, CT, ultrasound imaging, MRI, PET, and single photon emission computer tomography. MRI and CT have advantages over ultrasound imaging in the sense that higher resolution and clearer images are produced.

Imaging techniques have long been used for assessing and treating cardiac [67] and carotid disease [8, 24, 75]. Today's available imaging modalities produce a wide range of image data types for disease assessment, which include two-dimensional (2D) projection images, reconstructed three-dimensional (3D) images, 2D slice images, true 3D images, time sequences of 2D and 3D images, and sequences of 2D interior view (endoluminal) images. The use of ultrasound in the diagnosis and the assessment of imaging organs and soft tissue structures, as well as human blood, is well established [26] (see Figure 1.1, which illustrates two imaging scanners). Because of its noninvasive nature and continuing improvements in imaging quality, ultrasound imaging is progressively achieving an important role in the assessment and the characterization of cardiac imaging (see Figure 1.2), and the assessment of carotid artery disease [40, 53, 59, 60, 71, 75] (see Figure 1.3). The main disadvantage of ultrasound is that it does not work well in the presence of bone or gas, and the operator needs a high level of skill in both image acquisition and interpretation to carry out the clinical evaluation. On the other hand, standard angiography cannot give reliable information on the cross-sectional structure of the arteries [53]. This makes it difficult to accurately assess the build-up of plaque along the artery walls. B-mode ultrasound imaging or intravascular ultrasound (IVUS) has emerged, and it is widely used for visualizing carotid plaques and assessing plaque characteristics that are related to the onset of neurological symptoms. IVUS needs the insertion of a catheter into a vessel of interest that is equipped with an ultrasonic transducer enabling the reproduction of real-time cross-sectional images. However, reproducible measurements of the severity of the plaque in

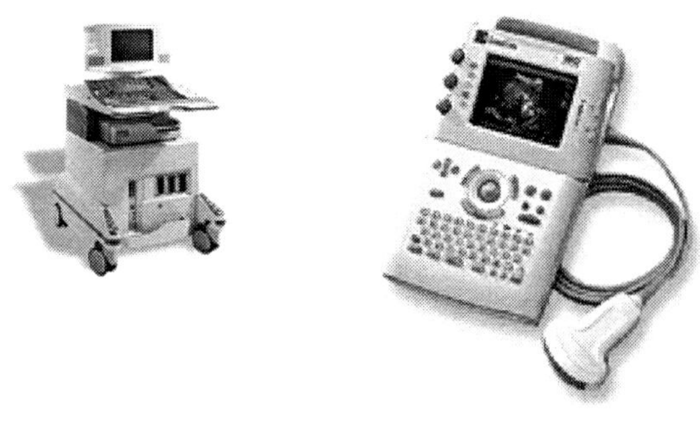

(a) (b)

FIGURE 1.1: Ultrasound imaging scanners: (a) ATL™ HDI-5000 and (b) ATL™ Somnosite 180 plus portable ultrasound system [84].

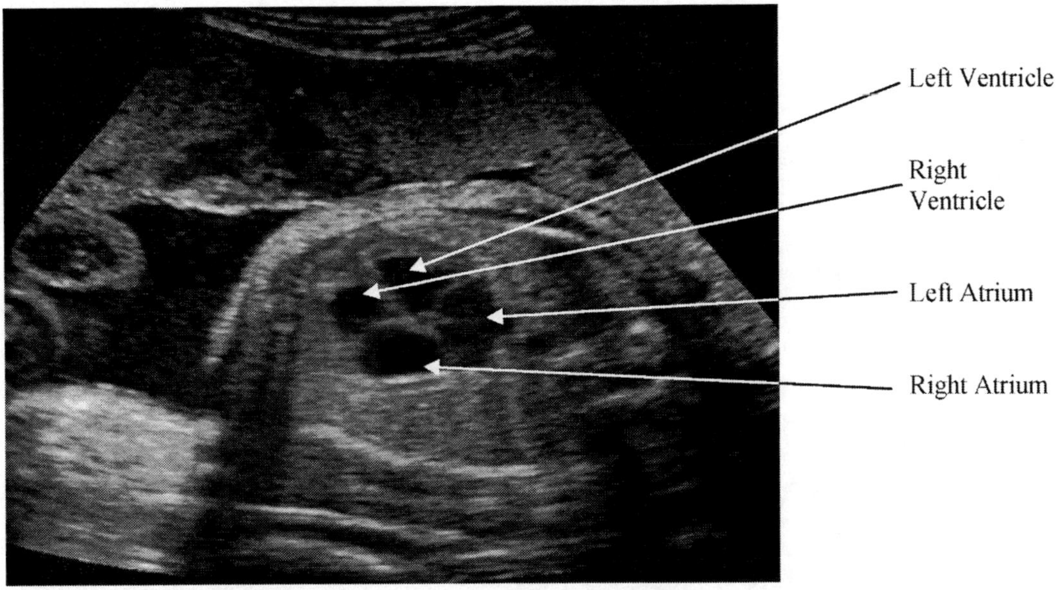

FIGURE 1.2: Ultrasound B-mode cardiac image, where the left ventricle (LV), the right ventricle (RV), the left atrium (LA), and the right atrium (RA) are indicated.

2D and 3D ultrasound are made difficult because of the complex shapes, the asymmetry of carotid plaques, and the speckle noise that is present in ultrasound images [26]. Furthermore, IVUS is an invasive method, as a catheter is inserted in the artery under investigation and possesses, therefore, a certain risk for the patient.

The use of ultrasound in medicine began during the Second World War in various centers around the world. The work of Dr. Karl Theodore Dussik in Austria in 1942 [85] on ultrasound transmission investigating the brain provides the first published work on medical ultrasonics. Furthermore, although other researchers in the United States, Japan, and Europe have also been cited as pioneers, the work of Prof. Ian Donald and his colleagues [86] in Glasgow, in the mid-1950s, did much to facilitate the development of practical ultrasound technology and applications. This led to the wider use of ultrasound in medical practice in subsequent decades.

From the mid-1960s onward, the advent of commercially available systems allowed the wider dissemination of the use of ultrasound. Rapid technological advances in electronics and piezoelectric materials provided further improvements from bistable to grayscale images and from still images to real-time moving images. The technical advances at this time (mid-1960s) led to the rapid growth in the applications of ultrasound. The development of Doppler ultrasound [87] has been progressing alongside the imaging technology, but the fusing of the two technologies in duplex scanning and the subsequent development of color Doppler imaging provided even more scope for investigating

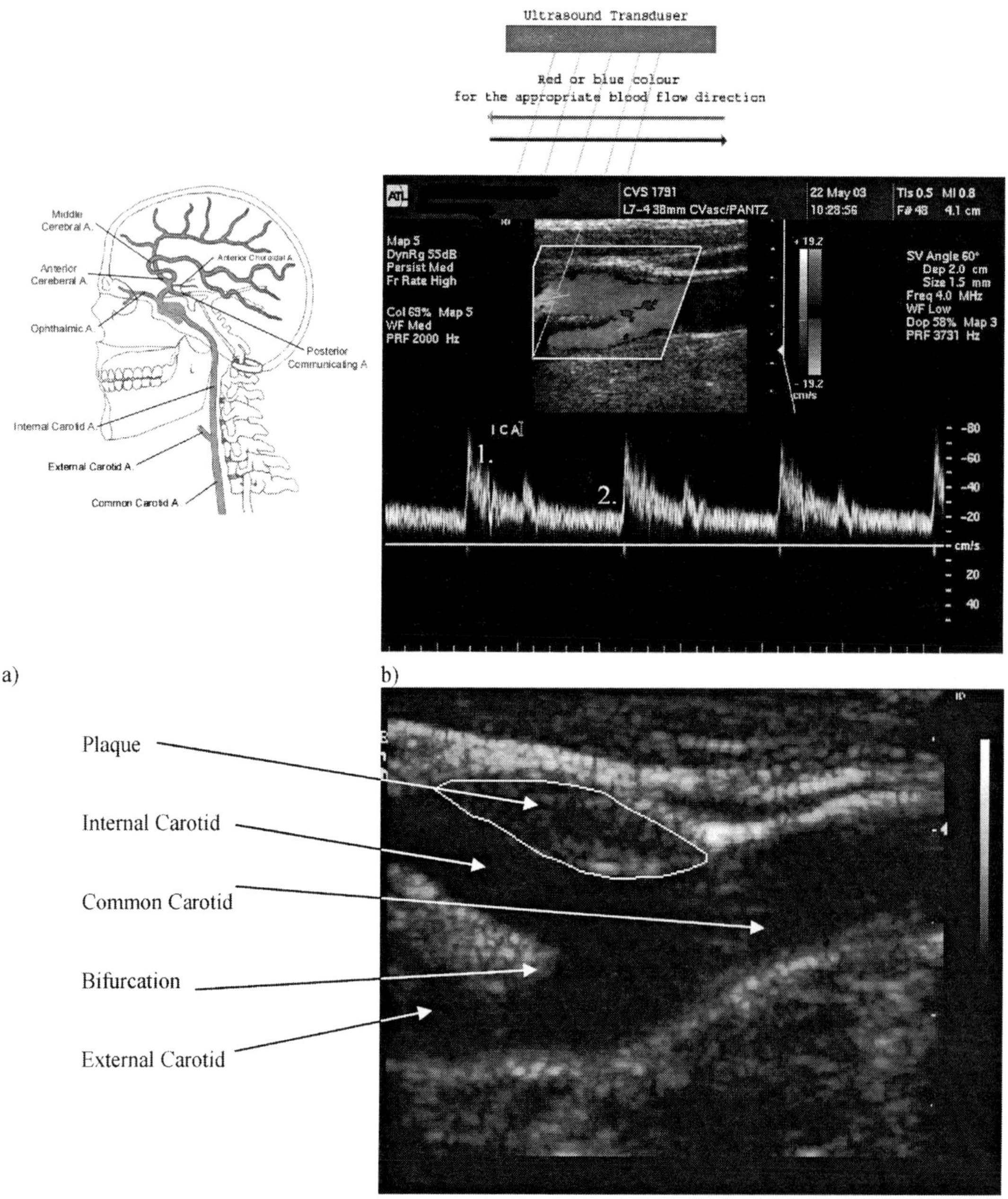

Ultrasound Transduser
Red or blue colour
for the appropriate blood flow direction
Middle Cerebral A.
Anterior Cerebral A.
Anterior Choroidal A.
Ophthalmic A.
Posterior Communicating A.
Internal Carotid A.
External Carotid A.
Common Carotid A.
CVS 1791
L7-4 38mm CVasc/PANTZ
22 May 03
10:28:56
TIs 0.5 MI 0.8
F# 48 4.1 cm
Map 5
DynRg 55dB
Persist Med
Fr Rate High
Col 69% Map 5
WF Med
PRF 2000 Hz
SV Angle 60°
Dep 2.0 cm
Size 1.5 mm
Freq 4.0 MHz
WF Low
Dop 58% Map 3
PRF 3731 Hz
+19.2
-19.2
cm/s
ICA
1.
2.
-80
-60
-40
-20
cm/s
20
40
Plaque
Internal Carotid
Common Carotid
Bifurcation
External Carotid

a)

b)

c)

circulation and blood supply to organs, tumors, etc. The advent of the microchip in the 1970s and the subsequent exponential increase in processing power facilitated the development of faster and more powerful systems incorporating digital beam forming, signal enhancement, and new ways of interpreting and displaying data, such as power Doppler [87] and 3D imaging [46]. Ultrasound has long been recognized as a powerful tool for use in the diagnosis and the evaluation of many clinical entities. Over the past decade, as higher quality and less expensive scanners are widely available, ultrasound has proliferated throughout various specialties.

1.1.1 Basic Principles of Ultrasound Imaging

Ultrasound is a sound wave with a frequency that exceeds 20 kHz. It transports energy and propagates through several means as a pulsating pressure wave. It is described by a number of wave parameters such as pressure density, propagation direction, and particle displacement. If the particle displacement is parallel to the propagation direction, then the wave is called a longitudinal or compression wave. If the particle displacement is perpendicular to the propagation direction, it is a shear or transverse wave. The interaction of ultrasound waves with tissue is subject to the laws of geometrical optics. It includes reflection, refraction, scattering, diffraction, interference, and absorption. Except from interference, all other interactions reduce the intensity of the ultrasound beam.

The main characteristic of an ultrasound wave is its wavelength , which is a measure of the distance between two adjacent maximum or minimum values of a sine curve and its frequency f, which is the number of waves per unit of time. The product of these two measures gives the velocity of ultrasound wave propagation described by the equation $=f$. Ultrasound techniques are mainly based on measuring the echoes transmitted back from a medium when sending an ultrasound wave to it. In the echo impulse ultrasound technique, the ultrasound wave interacts with tissue and blood, and some of the transmitted energy returns to the transducer to be detected by

---

FIGURE 1.3: (a) The carotid system illustrating the common carotid artery, its bifurcation, and the internal and external carotid arteries [110]. (b) Longitudinal color flow duplex image of the carotid artery combined with the Doppler ultrasound image. The highlighted image with white contour on top shows the carotid bifurcation. The 2D signal shows the velocity variation that is related to the cardiac cycle. A blood flow velocity spectrum is displayed with markings 1 and 2, where marking 1 represents the peak systolic velocity, and marking 2 represents the end diastolic velocity. This is the duration of one cardiac cycle. Different colors (shades) represent blood flow direction. For the current picture, red represents the blood moving to the brain through the carotid artery, whereas blue represents the blood returning back from the brain. (c) Ultrasound B-mode longitudinal image of the carotid bifurcation with a manually outlined plaque, which is usually confirmed with the blood flow image. (For interpretation of the reference to color in this figure legend, the reader is referred to the Web version of this book.)

the instrument. If we know the velocity of propagation in the tissue under investigation, we can determine the distance from the transducer at which the interaction occurs [88]. The characteristics of the return signal (amplitudes, phases, etc.) provide information on the nature of the interaction, and, hence, they give some indication of the type of the medium in which they occurred. Mainly two principles are used in medical ultrasound diagnostics—the echo impulse technique and the Doppler technique [88].

The second principle used in ultrasound diagnostics is the Doppler principle, named after the physicist Christian Doppler (1803–1853) [89]. This technique is based on the principle that the received frequency of sound echoes reflected by a moving target is related to the velocity of the target. The frequency shift (the Doppler frequency shift) f of the echo signal is proportional to the flow velocity (in centimeters per second) and the ultrasound transmission frequency f (in megahertz). The Doppler shift is described by the formula $f = 2f_0(\ \cos\ )/u_{sp}$, where f_0 is the transmitted frequency of the signal, is the angle between the direction of movement of the moving object and the ultrasound beam, and u_{sp} is the speed of sound through tissue that is approximately 1540 m/s.

In Doppler ultrasound, waves are produced by a vibrating crystal using the piezoelectric effect, whereas the returned echoes are displayed as a 2D signal, as shown in Figure 1.3b. When blood flow in a vessel is examined, sound reflections caused by the blood's corpuscular elements play a major role. Based on the fact that blood flow velocity varies in different areas of a vessel, the Doppler signal contains a broad frequency spectrum. In a normal internal carotid artery (ICA), the spectrum varies from 0.5 to 3.5 kHz, and is less than 120 cm/s when an ultrasound beam of 4 MHz is used.

1.1.2 Ultrasound Modes

The two main scanning modes are A- and B-modes. Other modes used are M-mode, duplex ultrasound, color-coded ultrasound, and power Doppler ultrasound, which will be briefly introduced below.

A-mode refers to amplitude mode scanning, which is mainly of historical interest. In this mode, the strength of the detected echo signal is measured and displayed as a continuous signal in one direction. A-mode is a line, with strong reflections being represented as an increase in the signal amplitude. This scanning technique has the limitation that the recorded signal is 1D with limited anatomical information. A-mode is no longer used, especially for the assessment of cardiovascular disease. Its use is restricted to very special uses such as in ophthalmology to perform very accurate measurements of distance.

B-mode refers to the brightness mode. In B-mode, echoes are displayed as a 2D grayscale image. The amplitude of the returning echoes is represented as dots (pixels) of an image with different gray values, as illustrated in Figure 1.3b and 1.3c. The image is constructed by these pixels line by line. Advances in B-mode ultrasound have resulted in improved anatomic definition, which has enabled plaque characterization [39, 88].

M-mode is used in cardiology, and it is actually an A-scan plotted against time. The result is the display of consecutive lines plotted against time. Using this mode, detailed information may be obtained about various cardiac dimensions and also the accurate timing of vascular motion.

Moving blood (see Figure 1.3b) generates a Doppler frequency shift in the reflected sound from insonated red blood cells, and this frequency shift can be used to calculate the velocity of the moving blood using the Doppler equation [89]. The invention of gated Doppler ultrasound in the late 1950s allowed velocity sampling at different depths and positions, and its subsequent combination with B-mode real-time ultrasonic imaging led to the development of duplex ultrasound. Stenosis in any vessel is characterized by an increase in systolic and diastolic velocities. Several types of Doppler systems are used in medical diagnosis: continuous wave (CW) Doppler, pulsed wave (PW) Doppler, duplex ultrasound, and color flow duplex (see also Figure 1.3b). In CW Doppler, the machine uses two piezoelectric elements serving as transmitters and receivers. They continuously transmit ultrasound beams. Because of the continuous way that ultrasound is being transmitted, no specific information about depth can be obtained. PW Doppler is used to detect blood flow at a specific depth. Sequences of pulses are transmitted to the human body, which are gated for a short period of time to receive the echoes. By selecting the time interval between the transmitted and received pulses, it is possible to examine vessels at a specific depth.

In color-coded ultrasound, every pixel is tested for Doppler shift. Using this technique, the movement of the red blood cells is finally depicted through color. The final image results by superimposing the color-coded image on the B-mode image.

Power Doppler is the depiction of flow based on the integrated power of the Doppler spectrum rather than on the mean Doppler frequency. This modality results in an angle, which is independent of the resulting enhanced sensitivity in flow detection as compared to the color-coded Doppler, and, therefore, the detection of low flow is better viewed.

1.1.3 Image Quality and Resolution

The quality of the produced ultrasound image depends on image resolution, axial and lateral. Resolution is defined as the smallest distance between two points at which they can be represented as distinct. Axial resolution refers to the ability of representing two points that lie along the direction of ultrasound propagation. It depends on the wavelength of the beam. In B-mode, ultrasound pulses consist of one to two sinusoidal wavelengths, and the axial resolution is dependent on the wavelength of the waveforms and lies in the range of the ultrasound wavelength (= 0.21 mm). Resolution depends on the frequency of the beam waveforms. Since this value is reciprocal to the ultrasound frequency (= /f), the axial resolution improves with increasing frequency.

Lateral resolution refers to the ability to represent two points that lie at a right angle to the direction of ultrasound propagation. This is dependent on the width of the ultrasound wave (beam). To be able to resolve points that lie close together, the width of the ultrasound beam has to be kept reasonably small, and the diameter of the transducer is kept as large as possible (i.e., small phase-array transducers have worse lateral resolution than large linear or curved-array transducers).

To achieve the best results in vascular ultrasound imaging, the transmission frequencies are in the range of 1–10 MHz. The selected frequency depends on the application domain. For arteries that are located close to the human skin, frequencies greater than 7.5 MHz are used, whereas for arteries that are located deeper in the human body, frequencies from 3 to 5 MHz are used. For transcranial applications, frequencies less than 2 MHz are used. However, when selecting a frequency, the user has to keep in mind that the axial resolution is proportional to the ultrasound wavelength, whereas the intensity of the signal depends on the attenuation of the signal transmitted through the body, with the higher the frequency, the higher the attenuation. Therefore, there is a tradeoff between higher resolution ultrasound images at smaller depths and lower resolution images at higher depths.

1.1.4 Limitations of Ultrasound Imaging

Variability in B-mode images (even when using the same ultrasonic equipment with fixed settings) does exist [75]. Sources of variability are outlined below.

1. Geometrical and diffraction effects, where spatial compound imaging may be employed to correct the image [39, 89].
2. Interpatient variation due to depth dependence and inhomogeneous intervening tissue, where normalization techniques may be applied to standardize the image [54, 55, 59, 60].
3. Speckle noise affecting the quality of ultrasound B-mode imaging. It is described as an ultrasound textural pattern that varies depending on the type of the biological tissue. The presence of speckle, which is difficult to suppress [5–28], may obscure small structures, thus degrading the spatial resolution of an ultrasonic image [59]. Despeckle filtering may be applied to improve the quality of the image.
4. Low contrast of the intima media complex or plaque borders [24, 53] and a small thin size [54, 60], making the image interpretation a difficult task.
5. Falsely low echogenicity due to shadowing effects, hindering the observation in B-mode images, of plaques or the intima media complex or other structures [53].
6. Low signal-to-noise ratio (SNR) in anechoic components and difficulty in outlining the carotid plaque, or other tissue under investigation, where the difficulty may be overcome by employing the use of color-coded images [54, 55].
7. Intraobserver variability where the ultrasound images inspected by the same expert at different occasions might be differently evaluated [59, 60].

8. Interobserver variability where the ultrasound images inspected by two or more experts might be differently evaluated [59].

It is noted that entries 7 and 8 are applicable in any medical imaging modality. To overcome intraobserver and interobserver variability, it is generally recommended that multiple observers should perform the image evaluation.

1.2 SPECKLE NOISE

In this section, we introduce speckle noise as a major factor limiting visual perception and processing of ultrasound [and synthetic aperture radar (SAR) images] [2–4, 9]. A mathematical speckle model for ultrasound images is introduced, where the statistics of speckle noise are presented, taking into consideration the log compression of the ultrasound image, which is performed to match the image into the display device (see Section 1.2.2). Based on this speckle model, a number of despeckling techniques are derived and explained in detail in Chapter 2. Specifically, the following categories of despeckle filtering techniques are presented: linear filtering (local statistics filtering, homogeneity filtering), nonlinear filtering (median filtering, linear scaling filtering, geometric filtering, logarithmic filtering, homomorphic filtering), anisotropic diffusion filtering (anisotropic diffusion, speckle-reducing anisotropic diffusion, coherent nonlinear anisotropic diffusion), and wavelet filtering.

Noise and artifacts can cause signal and image degradation for many medical image modalities. Different image modalities exhibit distinct types of degradation. Images formed with coherent energy, such as ultrasound, suffer from speckle noise. Image degradation can have a significant impact on image quality and, thus, affect human interpretation and the accuracy of computer-assisted methods. Poor image quality often makes feature extraction, analysis, recognition, and quantitative measurements problematic and unreliable. Therefore, image despeckling is a very important task, which motivated a significant number of studies in medical imaging [14, 22–24, 26, 28, 31].

The use of ultrasound in the diagnosis and the assessment of arterial disease is well established because of its noninvasive nature, its low cost, and the continuing improvements in image quality [1]. Speckle is a form of locally correlated multiplicative noise that corrupts medical ultrasound imaging making visual observation difficult [2, 3]. The presence of speckle noise in ultrasound images has been documented since the early 1970s, where researchers such as Burckhardt [2], Wagner et al. [3], and Goodman [4] described the fundamentals and the statistical properties of the speckle noise. Speckle is not truly noise in the typical engineering sense since its texture often carries useful information about the image being viewed [2–4].

Speckle noise is the primary factor that limits the contrast resolution in diagnostic ultrasound imaging, thereby limiting the detectability of small low-contrast lesions and making the ultrasound images generally difficult for the nonspecialist to interpret [2, 3, 5, 6]. Because of the speckle presence, ultrasound experts with sufficient experience may not often draw useful conclusions from the images [6]. Speckle also limits the effective application (e.g., edge detection) of automated computer-aided analysis (e.g., volume rendering, 3D display) algorithms. It is caused by the interference between ultrasound waves reflected from microscopic scattering through the tissue.

Therefore, speckle is most often considered a dominant source of noise in ultrasound imaging and should be filtered out [2, 5, 6] without affecting important features of the image. In this book, we carry out a comparative evaluation of despeckle filtering techniques based on texture analysis, image quality evaluation metrics, as well as visual assessment by experts on 440 ultrasound images of the carotid artery bifurcation. Results of this study were also published in Ref. [7]. Moreover, a comparative evaluation framework for the selection of the most appropriate despeckle filter for the problem under investigation is proposed.

1.2.1 Physical Properties and the Pattern of Speckle Noise

The speckle pattern, which is visible as the typical light and dark spots the image is composed of, results from destructive interference of ultrasound waves scattered from different sites. The nature of speckle has been a major subject of investigation [2–4, 12, 31]. When a fixed rigid object is scanned twice under exactly the same conditions, one obtains identical speckle patterns. Although of random appearance, speckle is not random in the same sense as electrical noise. However, if the same object is scanned under slightly different conditions, say, with a different transducer aperture, pulse length, or transducer angulation, the speckle patterns change.

The most popular model adopted in the literature to explain the effects that occur when a tissue is insonated is illustrated in Figure 1.4, where a tissue may be modeled as a sound absorbing medium containing scatterers, which scatter the sound waves [56, 83]. These scatterers arise from inhomogeneity and structures approximately equal to or smaller in size than the wavelength of the ultrasound, such as tissue parenchyma, where there are changes in acoustic impedance over a microscopic level within the tissue. Tissue particles that are relatively small in relation to the wavelength (i.e., blood cells), and particles with differing impedance that lie very close to one another, cause scattering or speckling. Absorption of the ultrasound tissue is an additional factor to scattering and refraction, responsible for pulse energy loss. The process of energy loss involving absorption, reflection, and scattering is referred to as attenuation, which increases with depth and frequency. Because a higher frequency of ultrasound results in increased absorption, the consequence is a decrease in the depth of visualization.

Figure 1.5 illustrates the entire scattering procedure [56]. Consider a transducer insonating a homogeneous medium containing four point-like scatterers, as depicted in Figure 1.5a. These scatterers yield spherical waves that will arrive at the transducer at slightly different times after the transmission of the ultrasound pulse. Usually, the pulse envelope is approximately Gaussian,

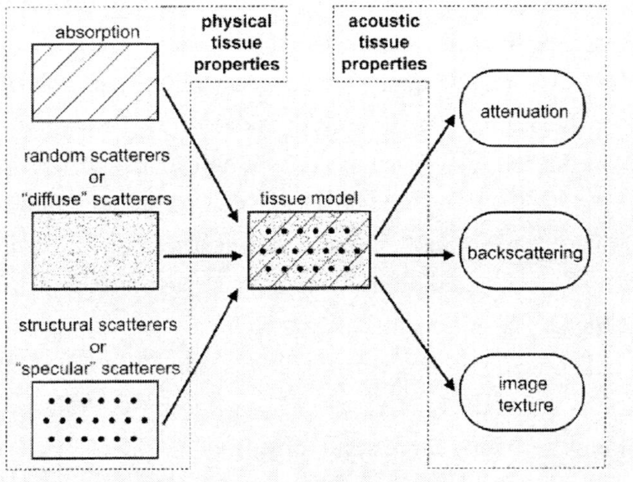

FIGURE 1.4: The usual tissue model in ultrasound imaging (modified from Ref. [56]).

as shown in Figure 1.5b. If the pulse has a Gaussian shape, then so is its spectrum. One chooses a Gaussian shape because for a medium with a linear attenuation coefficient, this Gaussian shape of the spectrum is maintained while the pulse travels through the medium (although a shift of this Gaussian spectrum to lower frequencies occurs while the pulse travels through the medium because the attenuation increases with the frequency).

Upon reception of the reflected signal, the transducer produces an electrical signal [radio frequency (RF)] that is the algebraic sum of the instantaneous sound pressures originating from the backscattered waves (four waves in Figure 1.5a). The depth differences of the scatterers are smaller than the axial size of the resolution volume of the transducer (i.e., the pulse length). This is, in fact,

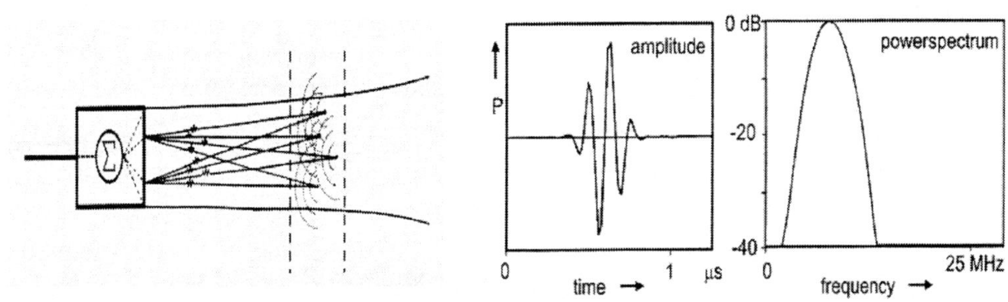

(a) (b)

FIGURE 1.5: (a) The scattering in the sound beam. (b) One pulse in the time and frequency domains (from Ref. [56]).

the basic cause for the generation of tissue texture. The formed pattern is the so-called speckle pattern. Note, in particular, that the tissue texture resulting from this speckle pattern is, in general, not a true image of the histological structure of the tissue, but rather an interference pattern that is mainly determined by the beam characteristics. Speckle is described as one of the more complex image noise models [3, 4, 31, 56]; it is signal dependent, non-Gaussian, and spatially dependent.

In homogeneous tissue, the distribution of the scatterers throughout the 3D space is assumed to be isotropic. As displayed in Figure 1.4, one distinguishes random (or diffuse) scatterers and structural (or specular) scatterers. The diffuse scatterers are assumed to be uniformly distributed over space. Diffuse scattering arises when there are a number of scatterers with a random phase within the resolution cell of the ultrasound beam. This random nature of the location of the scatterers causes the statistical nature of the echo signals and, hence, the resulting speckle pattern. Consequently, a statistical approach to its analysis seems obvious.

Other properties of the tissue that affect the ultrasound as it propagates through it are the propagation speed, the attenuation, and the backscattering. The absorption of ultrasound is caused by relaxation phenomena of biological macromolecules [58] that transfer mechanical energy into heat. Another source of attenuation is the scattering, i.e., omnidirectional reflections by small inhomogeneity in the tissue. The overall attenuation is, therefore, the result of absorption and scattering (as illustrated in Figure 1.4), which are both frequency dependent in such a way that the attenuation increases with frequency.

In analyzing speckle, an important point to bear in mind is to make a clear distinction between the speckle as it appears in the image and the speckle in the received RF signal. The block diagram in Figure 1.6 explains the entire track of the RF signal from the transducer to the screen inside the ultrasound imaging system. As set forth, the signal is subject to several transformations that severely affect its statistics. The most important of these is the log compression of the signal, which is employed to reduce the dynamic range of the input signal to match the lower dynamic range of the display device. The input signal could have a dynamic range of the order of 50–70 dB, whereas a typical display could have a dynamic range of the order of 20–30 dB. Such a relation is normally affected through an amplifier, which has a reducing amplification for a larger input signal.

In addition, the expert has the possibility to manually adjust several machine settings. In Figure 1.6, these are indicated as the slide contact overall gain and the time gain compensation (TGC). These machine settings control the amplification of the signal—the overall gain controls the overall amplification, and the TGC is a time-dependent amplification—and serve as tools for the expert to adjust the image for an optimal visual diagnosis. The TGC is adjusted by several (usually seven) slide contacts, each of which controls the gain in part of the image. For instance, if the slide contacts are placed in a vertical row, the top slide contact controls the gain in the top of the image, the bottom slide contact controls the gain in the bottom of the image, etc. This position-specific gain in

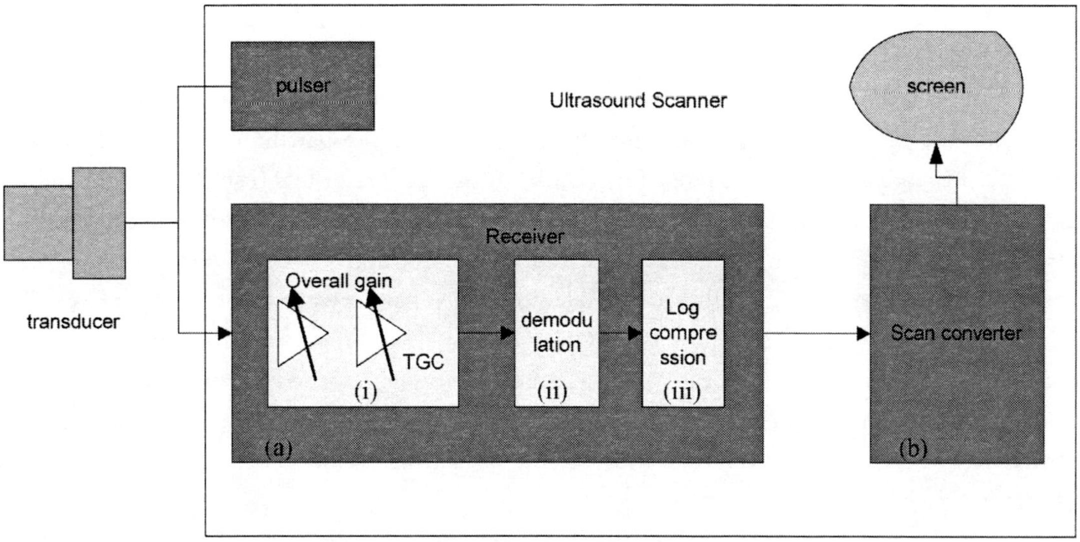

FIGURE 1.6: The processing steps of the RF signal inside the ultrasound scanner (modified from Ref. [57]).

the image is realized by making the amplification of the signal dependent on the exact time that the sound reflection is received. Since the position where a pixel is displayed on the screen is dependent on this time instant, the time-dependent amplification of the received signal converts to a position-dependent change in the gray value of the pixels on the screen.

1.2.2 Speckle Noise Modeling

To be able to derive an efficient despeckle filter, a speckle noise model is needed. The speckle noise model for both ultrasound and SAR images may be approximated as multiplicative [31]. The signal at the output of the receiver demodulation module of the ultrasound imaging system [see Figure 1.6a(ii)] may be defined as

$$y_{i,j} = x_{i,j} n_{i,j} + a_{i,j}, \qquad (1.1)$$

where $y_{i,j}$ represents the noisy pixel in the middle of the moving window, $x_{i,j}$ represents the noise-free pixel, $n_{i,j}$ and $a_{i,j}$ represent the multiplicative and additive noise, respectively, and i,j are the indices of the spatial locations that belong in the 2D space of real numbers, $i,j \quad \Re^2$.

Despeckling is based on estimating the true intensity $x_{i,j}$ as a function of the intensity of the pixel $y_{i,j}$ and some local statistics calculated on a neighborhood of this pixel.

Wagner et al. [3] showed that the histogram of amplitudes within the resolution cells of the envelope-detected RF signal backscattered from a uniform area with a sufficiently high scatterer density has a Rayleigh distribution with mean proportional to the standard deviation (with / = 1.91). This implies that speckle could be modeled as multiplicative noise.

However, the signal processing stages inside the scanner modify the statistics of the original signal, i.e., the logarithmic compression [see Figure 1.6a(iii)]. The logarithmic compression is used to adjust the large echo dynamic range (50–70 dB) to the number of bits (usually 8) of the digitization module in the scan converter (see Figure 1.6b). More specifically, logarithmic compression affects the high-intensity tail of the Rayleigh and Rician probability density functions more than the low-intensity part. As a result, the speckle noise becomes very close to the white Gaussian noise corresponding to the uncompressed Rayleigh signal [31]. In particular, it should be noted that speckle is no longer multiplicative in the sense that, on homogeneous regions, where $x_{i,j}$ can be assumed constant, the mean is proportional to the variance (^{2}) rather than the standard deviation ( ) [24, 26, 28, 31]. In this respect, the speckle index *C* will be for the log-compressed ultrasound images, i.e., $C = {}^2/$.

Referring back to Eq. (1.1), since the effect of the additive noise is considerably smaller compared with that of the multiplicative noise, it may be written as

$$y_{i,j} \quad x_{i,j} n_{i,j}. \tag{1.2}$$

Thus, the logarithmic compression transforms the model in Eq. (1.2) into the classical signal in the additive noise form as

$$\log(y_{i,j}) = \log(x_{i,j}) + \log(n_{i,j}) \tag{1.3}$$

and

$$g_{i,j} = f_{i,j} + nl_{i,j}. \tag{1.4}$$

For the rest of the book, the term $\log(y_{i,j})$, which is the observed pixel on the ultrasound image display after logarithmic compression, is denoted as $g_{i,j}$, and the terms $\log(x_{i,j})$ and $\log(n_{i,j})$, which are the noise-free pixel and the noise component after logarithmic compression, are denoted as $f_{i,j}$ and $nl_{i,j}$, respectively [see Eq. (1.4)].

1.2.3 Early Attempts of Despeckle Filtering in Different Modalities and Ultrasound Imaging

The widespread of ultrasound imaging equipment, including mobile and portable telemedicine ultrasound scanning instruments and computer-aided systems, necessitates the need for better image processing techniques to offer a clearer image to the medical practitioner. This makes the use of efficient despeckle filtering a very important task. Early attempts to suppress speckle noise were implemented by averaging of uncorrelated images of the same tissue recorded under different spatial positions [5, 9, 10]. Although these methods are effective for speckle reduction, they require multiple images of the same object to be obtained [11]. Speckle-reducing filters originated from the SAR community [9]. These filters then have later been applied to ultrasound imaging since the early 1980s [12]. Filters that are widely used in both SAR and ultrasound imaging were originally proposed by Lee [9, 14, 15], Kuan et al. [11], Frost et al. [13], and Kuan and Sawchuk [16].

Some researchers have tried in the past to despeckle SAR images by averaging of uncorrelated images obtained from different spatial positions [46]. These temporal averaging and multiframe methods aimed to increase the SNR by generating multiple uncorrelated images that are incoherently summed to reduce speckle [82]. Despite being simple and fast, these approaches suffer from two limitations. First, to produce uncorrelated ultrasound images, the transducer has to be translated at least by about half its element width for each of the generated frames [2]. Second, temporal averaging based on transducer movement causes the loss of small details such as small vessels and texture patterns because of blurring. For the above reasons, this procedure has been proven to be not suitable for despeckle filtering. It is most suitable for additive noise reduction [46, 82]. Another disadvantage of this method is that multiple images from the same object are required [10, 15]. Other researchers applied their techniques on the ultrasound images of the kidney [26], echocardiograms [27], heart [24], abdomen [24], pig heart [28], and liver [63], on SAR images [17, 34, 77, 78], and on real-world [16, 28] and artificial images [10, 66]. They used statistical measures, like the mean, the variance, the median, the speckle index (C), the mean-square error (MSE), the image contrast, and the visual perception evaluation made by experts, to evaluate their techniques. They compared their despeckling techniques with the Lee filter [9], homomorphic filtering [17, 18], the median filter [33], and diffusion filtering [5, 20–23]. Despeckle filtering can also be used as a preprocessing step for image segmentation [7, 54, 59, 60] or image registration [46] techniques. By suppressing the speckle, the performance of these techniques can be improved.

Many authors have shown a reduction of lesion detectability of approximately a factor of 8 due to the presence of speckle noise in the image [2, 4, 13]. This radical reduction in contrast resolution is responsible for the poorer effective resolution of ultrasound compared to X-ray and MRI [46]. Despeckle filtering is, therefore, a critical preprocessing step in medical ultrasound images provided that the features of interest for the diagnosis are not lost.

1.3 AN OVERVIEW OF DESPECKLE FILTERING TECHNIQUES

Table 1.1 summarizes the despeckle filtering techniques for ultrasound imaging that are presented in this book, grouped under the following categories: linear filtering (local statistics filtering, homogeneity filtering), nonlinear filtering (median filtering, linear scaling filtering, geometric filtering,

TABLE 1.1: An overview of despeckle filtering techniques

SPECKLE REDUCTION TECHNIQUE	METHOD	INVESTIGATOR	FILTER NAME
Linear filtering	Moving window utilizing local statistics		
	1. Mean ( ) and variance (^{2})	[9–14] and [13–16]	*DsFlsmv*
	2. Mean, variance, third and fourth moments (higher statistical moments) and entropy	[9–14]	*DsFlsmvsk1d* *DsFlsmvsk2d*
	3. Homogeneous mask area filters	[32]	*DsFlsminsc*
	4. Wiener filtering	[2–15]	*DsFwiener*
Nonlinear filtering	Median filtering	[33]	*DsFmedian*
	Linear scaling of the gray-level values	[46]	*DsFls* *DsFca* *DsFlecasort*
	Based on the most homogeneous neighborhood around each pixel	[8]	*DsFhomog*
	Nonlinear iterative algorithm (geometric filtering)	[10]	*DsFgf4d*
	The image is logarithmically transformed, the FFT is computed and denoised, and the inverse FFT is computed and finally exponentially transformed back	[2, 17, 18]	*DsFhomo*
Diffusion filtering	Nonlinear filtering technique for simultaneously performing contrast enhancement and noise reduction	[2, 5, 12, 13, 19–23]	*DsFad*
	Exponential damp kernel filters utilizing diffusion	[5]	

TABLE 1.1: (*continued*)

SPECKLE REDUCTION TECHNIQUE	METHOD	INVESTIGATOR	FILTER NAME
	Speckle-reducing anisotropic diffusion based on the coefficient of variation	[24]	*DsFsrad*
	Coherence enhancing diffusion	[24]	*DsFnldif*
Wavelet filtering	Only the useful wavelet coefficients are utilized	[15, 25–29, 35]	*DsFwaveltc*

logarithmic filtering, homomorphic filtering), anisotropic diffusion filtering (anisotropic diffusion, speckle-reducing anisotropic diffusion, coherent nonlinear anisotropic diffusion), and wavelet filtering. Furthermore, in Table 1.1, the methodology used, the main investigators, and the corresponding filter names are given. These filters are briefly introduced in this chapter and presented in detail in Chapter 2.

Some of the linear filters are Lee [9, 14, 15], Frost [13], and Kuan [11, 16]. The Lee and Frost filters have the same structure, whereas the Kuan filter is a generalization of the Lee filter. Both filters form the output image by computing the central pixel intensity inside a filter-moving window, which is calculated from the average intensity values of the pixels and a coefficient of variation inside the moving window. Kuan considered a multiplicative speckle model and designed a linear filter based on the minimum-mean-square error criterion that has optimal performance when the histogram of the image intensity is Gaussian distributed. The Lee filter [9] is a particular case of the Kuan filter based on a linear approximation made for the multiplicative noise model. The Frost filter [13] makes a balance between the averaging and the all-pass filters. It was designed as an adaptive Wiener filter that assumed an autoregressive exponential model for the image.

In the nonlinear filtering group, the gray-level values are linearly scaled to despeckle the image [61]. Some of the nonlinear filters are based on the most homogeneous neighborhood around each image pixel [8]. Geometric filters [10] are based on nonlinear iterative algorithms, which increment or decrement the pixel values in a neighborhood based on their relative values. The method of homomorphic filtering [17, 18] is similar to the logarithmic point operations used in histogram enhancement, where dominant bright pixels are de-emphasized. In the homomorphic filtering, the fast Fourier transform (FFT) of the image is calculated and then denoised, and then the inverse FFT is calculated.

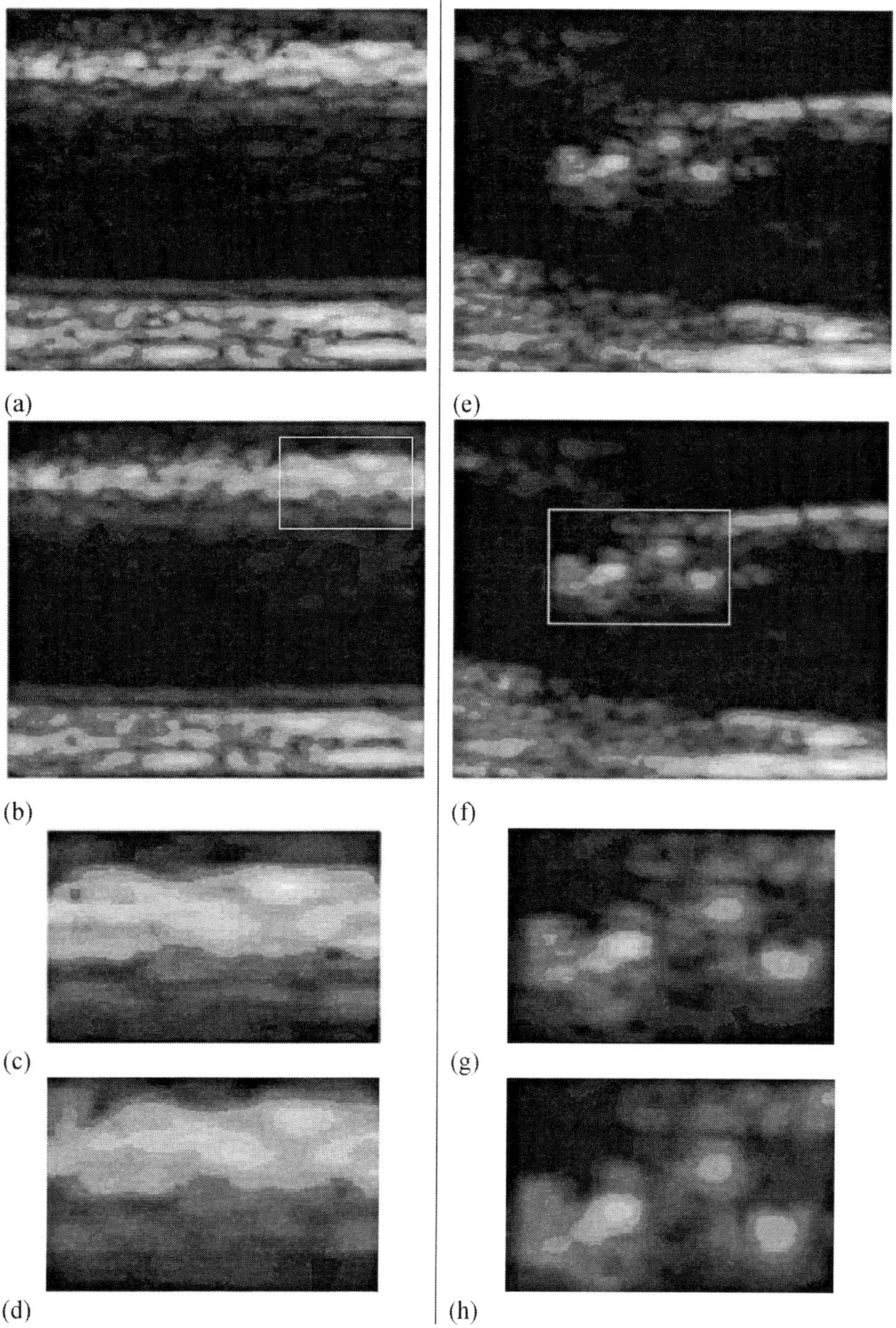
(a)
(e)
(b)
(f)
(c)
(g)
(d)
(h)

The diffusion filtering category includes filters based on anisotropic diffusion [2, 19, 20–23], coherence anisotropic diffusion [24], and speckle-reducing anisotropic diffusion [5]. These filters have been recently presented in the literature and are nonlinear filtering techniques. They simultaneously perform contrast enhancement and noise reduction by utilizing the coefficient of variation [5]. Furthermore, in the wavelet category, filters for suppressing the speckle noise were documented. These filters are making use of a realistic distribution of the wavelet coefficients [2, 15, 25–30], where only the useful wavelet coefficients are utilized. Different wavelet shrinkage approaches were extensively investigated based on Donoho's work [29].

Figure 1.7 illustrates original longitudinal asymptomatic (see Figure 1.7a) and symptomatic images (see Figure 1.7e) and their despeckled images (see Figure 1.7b and 1.7f). Asymptomatic images were recorded from patients at risk of atherosclerosis in the absence of clinical symptoms, whereas symptomatic images were recorded from patients at risk of atherosclerosis, which have already developed clinical symptoms, such as a stroke episode. Figure 1.7c–1.7h shows an enlarged window from the original and despeckled images (shown in a rectangle in Figure 1.7b and 1.7f).

1.4 LIMITATIONS OF DESPECKLE FILTERING TECHNIQUES

Despeckling is always a tradeoff between noise suppression and loss of information, which is something that experts are very concerned about. It is, therefore, desirable to keep as much important information as possible. The majority of speckle reduction techniques have certain limitations that can be briefly summarized as follows.

1. They are sensitive to the size and the shape of the window. The use of different window sizes greatly affects the quality of the processed images. If the window is too large, oversmoothing will occur, subtle details of the image will be lost in the filtering process, and edges will be blurred. On the other hand, a small window will decrease the smoothing capability of the filter and will not reduce the speckle noise, thus making the filter not effective. In homogenous areas, the larger the window size, the more efficient the filter in reducing the speckle noise. In heterogeneous areas, the smaller the window size, the more it is possible to keep subtle image details unchanged. Our experiments showed that a [7 × 7] window size is a fairly good choice.
2. Some of the despeckle methods based on window approaches require thresholds to be

---

FIGURE 1.7: Results of despeckle filtering based on linear filtering (first-order local statistics, *DsFlsmv*). Asymptomatic case: (a) original, (b) despeckled, (c) enlarged region marked in (b) of the original, and (d) enlarged region marked in (b) of the despeckled image. Symptomatic case: (e) original, (f) despeckled, (g) enlarged region marked in (f) of the original, (h) enlarged region marked in (f) of the despeckled image. Regions were enlarged by a factor of 3.

used in the filtering process, which have to be empirically estimated. There are a number of thresholds introduced in the literature, which include gradient thresholding [5], soft or hard thresholds [29], nonlinear thresholds [28], and wavelet thresholds [25, 28, 37]. The inappropriate choice of a threshold may lead to average filtering and noisy boundaries, thus leaving the sharp features unfiltered [7, 10, 14].

3. Most of the existing despeckle filters do not enhance the edges, but they only inhibit smoothing near the edges. When an edge is contained in the filtering window, the coefficient of variation will be high, and smoothing will be inhibited. Therefore, speckle in the neighborhood of an edge will remain after filtering. They are not directional in the sense that in the presence of an edge, all smoothing is precluded. Instead of inhibiting smoothing in directions perpendicular to the edge, smoothing in directions parallel to the edge is allowed.
4. Different evaluation criteria for evaluating the performance of despeckle filtering are used by different studies. Although most of the studies use quantitative criteria like the MSE and the speckle index (C), there are additional quantitative criteria like texture analysis and classification, image quality evaluation metrics, and visual assessment by experts that could be investigated.

1.5 GUIDE TO BOOK CONTENTS

In the following chapter, the theoretical background (equations), the algorithmic steps, and the MATLAB™ code of despeckle filters given in Table 1.1 are presented. Chapter 3 covers the evaluation methodology, the material and recording of ultrasound images, the texture and statistical analysis, the statistical k-nearest-neighbor classifier, the image quality evaluation metrics, and the experiments carried out for visual evaluation. Chapter 4 presents the applications of despeckle filtering techniques in ultrasound images of the carotid artery and in cardiac ultrasound images. Chapter 5 discusses, compares, and evaluates the proposed despeckle filtering techniques, where the strong and weak points for each filtering technique are presented. Reference is also made to video despeckling, where a video despeckling protocol is presented. Chapter 6 presents the summary and future directions, where a despeckling filtering and evaluation protocol is also proposed.

Finally, at the end of this book, an appendix provides additional information of all the functions used in this book, as these will be introduced in Chapter 2 and Chapter 3, together with the MATLAB™ code. Section A.2 of the appendix illustrates examples in MATLAB code for running the despeckle filtering toolbox functions.

• • • •

CHAPTER 2

Despeckle Filtering Algorithms

This chapter provides the basic theoretical background of despeckle filtering techniques together with their algorithmic implementation and MATLAB™ code for selected filters. The four groups of filters presented are linear filtering, nonlinear filtering, diffusion filtering, and wavelet filtering.

2.1 LINEAR FILTERING

Most of the techniques for speckle reduction filtering in the literature use linear filtering based on local statistics. Their working principle may be described by a weighted average calculation using subregion statistics to estimate statistical measures over different pixel windows varying from [3 × 3] up to [15 × 15]. All these techniques assume that the speckle noise model has a multiplicative form as given in Eq. (1.3) [7–15, 24, 26].

2.1.1 First-Order Statistics Filtering (*DsFlsmv* and *DsFwiener*)

The filters utilizing the first-order statistics such as the variance and the mean of the neighborhood may be described with the model as in Eq. (1.4). Hence, the algorithms in this class may be traced back to the following equation [5, 7–16] (see also Figure 2.1):

$$f_{i,j} = \bar{g} + k_{i,j}(g_{i,j} \quad \bar{g}), \tag{2.1}$$

where $f_{i,j}$ is the estimated noise-free pixel value, $g_{i,j}$ is the noisy pixel value in the moving window, $\bar{g}$ is the local mean value of an $N_1 \times N_2$ region surrounding and including pixel $g_{i,j}$, $k_{i,j}$ is a weighting factor, with k [0,1], and i and j are the pixel coordinates. The factor $k_{i,j}$ is a function of the local statistics in a moving window. It can be found in the literature [7, 9, 11, 14] and may be derived in different forms that

$$k_{i,j} = (1 \quad \bar{g}^2 \; ^2)/(\; ^2(1 + \; _n^2)) \tag{2.2}$$

$$k_{i,j} = \; ^2/(\bar{g}^2 \; _n^2 + \; ^2) \tag{2.3}$$

$$k_{i,j} = (\; ^2 \quad _n^2)/ \; ^2. \tag{2.4}$$

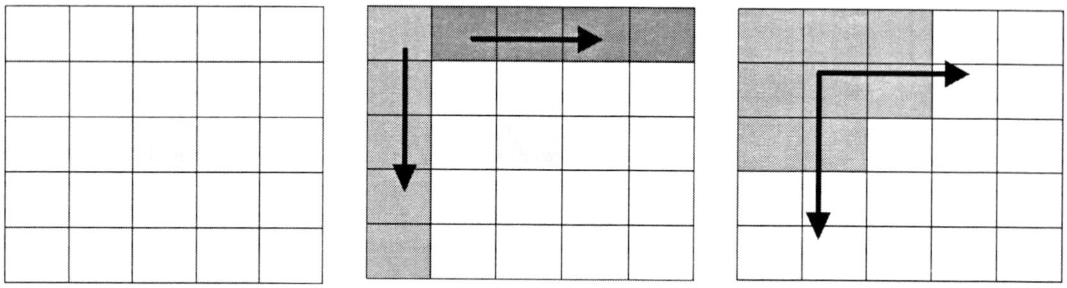

(a) (b) (c)

FIGURE 2.1: (a) Pixel moving window of [5 × 5] pixels. (b) Schematic operation of the filters *DsFlsminv1d* with a 1D sliding moving window. (c) *DsFlsmv* with a 2D sliding moving window.

The values ^{2} and $_{n}^{2}$ represent the variance in the moving window and the variance of noise in the whole image, respectively. The noise variance may be calculated for the logarithmically compressed image by computing the average noise variance over a number of windows with dimensions considerably larger than the filtering window. In each window, the noise variance is computed as

$$ {}_{n}^{2} = \overset{p}{\underset{i=1}{}} \; {}_{p}^{2}/\,\overline{g}_{p}, \qquad (2.5)$$

where $_{p}^{2}$ and $\overline{g}_{p}$ are the variance and the mean of the noise in the selected windows, respectively, and p is the index covering all windows in the whole image [24, 25, 31]. If the value of $k_{i,j}$ is 1 (in edge areas), this will result to an unchanged pixel, whereas a value of 0 (in uniform areas) replaces the actual pixel by the local average $\overline{g}$ over a small region of interest [see Eq. (2.1)]. In this study, the filter *DsFlsmv* uses Eq. (2.2).

The filter *DsFwiener* uses a pixel-wise adaptive wiener method [2–6, 13], which is implemented as given in Eq. (2.1), with the weighting factor $k_{i,j}$ as given in Eq. (2.4).

For both despeckle filters *DsFlsmv* and *DsFwiener*, the moving window size was [5 × 5] pixels (see also Figure 2.1a).

ALGORITHM 2.1

Linear filtering: linear scaling filter (*DsFlsmv*)

1. Load the image for filtering.
2. Specify the region of interest to be filtered, the moving window size (nhood), and the number of iterations (n).
3. Compute the noise variance $_{n}^{2}$ with Eq. (2.5) for the whole image.

4 Starting from the left upper corner of the image, compute for each moving window the coefficient $k_{i,j}$ in Eq. (2.2).

5 Compute $f_{i,j}$ in Eq. (2.1) and replace the noisy middle point in each moving window $g_{i,j}$ with the new computed value $f_{i,j}$.

6 Repeat steps 4 and 5 for the whole image by sliding the moving window from left to right.

7 Repeat steps 3–6 for *n* iterations.

8 Compute the image quality evaluation metrics and the texture features for the original and despeckled images.

9 Display the original and despeckled images, the image quality and evaluation metrics, and the texture features.

CODE 2.1

MATLAB code linear filtering: linear scaling filter (*DsFlsmv*)

```
function f = DsFlsmv(g, nhood, niterations)
% Despeckle filtering toolbox, © Christos P. Loizou 2007
%*************************************************************************
% Local first order statistics filter
% Input:
% g:              Original (input) noisy image
% nhood:          Size of the sliding moving window in pixels
% Iterations:     Number of iterations for which filtering is applied
%
% Output:
% f:              Despeckled (output) image
1 % Example: f=DsFlsmv (g, [5 5], 5);
% Load the image for filtering
%*************************************************************************
2 *
imshow (g);                 % show the original image
% Crop image region and select an area of interest to be despeckled
% Specify the area of interest to be filtered, the moving window size (nhood) and the number of
% iterations (n) the filtering is applied to the image
[x, y, BW, xi, yi]=roipoly(g);
maxx=max(xi); minx=min(xi); maxy=max(yi); miny=min(yi);
```

```
[xsize, ysize, imagec, rect]=imcrop(g, [minx miny (maxx-minx) (maxy-miny)]);
% the cropped image is g = imagec
% Check if the image loaded is a grayscale and normalize its values
if isa(imagec, 'uint8')
  u8out = 1;
  if (islogical(imagec))
      logicalOut = 1;
    imagec = double(imagec);
else
    logicalOut = 0;
    imagec = double(imagec)/255;
end
else
  u8out = 0;
end
% Calculate the noise and the standard deviation of the original image, and the noise variance
3 in the % whole image
% Compute the noise variance ^2_n with (2.5) from the whole image
stdnoise=(std2(imagec).*std2(imagec))/mean2(imagec);
noisevar=stdnoise*stdnoise; %noise variance
% Initialize a new image f (new image after filtering) with zeros
f = imagec;
% Apply n-iterations of the algorithm to the image
for i = 1:niterations
  fprintf('\rIteration %d',i);
  if i >=2
     imagec=f;
  end
% For each moving window, estimate the local mean of f.
localMean = filter2(ones(nhood), imagec) / prod(nhood);
% square of the local mean
4 lmsqr = localMean.*localMean;
% Starting from the left upper corner of the image, compute for each moving window the
% coefficient k_{i,j} in (2.2)
localVar = filter2(ones(nhood), imagec.^2) / prod(nhood) - localMean.^2;
% Compute f_{i,j} in (2.1) and replace the noisy middle point in each moving window g_{i,j},
5 with the % new computed value f_{i,j}
```

```
f=localMean + (localVar - lmsqr .*noisevar ./ max(0.1, localVar + lmsqr .* noisevar)) .*
(imagec - localMean);
end
% End for i Itterations
fprintf('\n');
if u8out==1,
  if (logicalOut)
    f = uint8(f);
else
6     f = uint8(round(f*255));
end
7 end
8 % Repeat steps 4 and 5 for for the whole image by sliding the moving window from left to
right
% Repeat steps 3 to 6 for n iterations specified
% Compute the texture and image quality evaluation metrics and display both the original
and the % despeckled images on the screen
% Calculate 61 Texture Features from the original and despeckled images
A=[ ]; F1=[ ] ;
%  Initialize the matrcies for texture features
T= DsTtexfeat(double(imagec));
A=[A, T'];
% Save the texture features of the original image in a matrix A
save or_texfeats A;

TAM=DsTtexfeat(double(f));
F1=[F1,TAM'];
save speckle1texfs F1;
9 % The texture features of the despeckled image are saved in matrix F1
% Call the function metrics to calculate and display the 19 different image quality metrics
between %  the original and the despeckled image
M=DsQmetrics(f, imagec);

% Show original and despeckled images on the screen
figure, imshow(imagec);
figure, imshow(f);
```

Algorithm 2.1 presents the algorithmic steps for the implementation of the *DsFlsmv* despeckle filter, whereas Code 2.1 presents the implementation of the filter in MATLAB™ code.

Figure 4.1b shows the application of the despeckle filter *DsFlsmv* on a phantom carotid artery image for a moving window size of [7 × 7] pixels and five iterations, whereas Figures 4.2 and 4.3 show the application of the filter with different window sizes and an increasing number of iterations, respectively. Figure 4.4b shows the results of the filter on an artificial carotid image (acquired by the ATL 5000 ultrasound scanner [7, 59, 84]), whereas Figures 4.6b and 4.7c and 4.7g present results on real carotid and cardiac ultrasound images, respectively. Finally, Figure 4.10 presents the results of the *DsFlsmv* despeckle filter applied on consecutive video frames. The number of iterations, as well as the size of the sliding moving window, is also shown.

Algorithm 2.2 presents the algorithmic steps for the implementation of the *DsFwiener* despeckle filter.

ALGORITHM 2.2

Linear filtering: linear scaling filter (*DsFwiener*)	
1	Load the image for filtering.
2	Specify the region of interest to be filtered, the moving sliding window size, and the number of iterations.
3	Compute the noise variance ^2_n with Eq. (2.5) for the whole image.
4	Starting from the left upper corner of the image, compute for each sliding moving window the coefficient $k_{i,j}$ in Eq. (2.4).
5	Compute $f_{i,j}$ in Eq. (2.1) and replace the noisy middle point in each moving window $g_{i,j}$ with the new computed value $f_{i,j}$.
6	Repeat steps 4 and 5 for all the pixels in the image by sliding the moving window from left to right.
7	Repeat steps 3–6 for a second iteration of despeckle filtering.
8	Compute the image quality evaluation metrics and the texture features for the original and despeckled images.
9	Display the original and despeckled images, the image quality and evaluation metrics, and the texture features.

Figure 4.1d shows the application of the despeckle filter *DsFwiener* on a phantom carotid artery image for a moving window size of [5 × 5] pixels and four iterations. Figure 4.4d shows the

results of the filter on an artificial carotid image, whereas Figure 4.6d presents the results on a real carotid ultrasound image.

2.1.2 Local Statistics Filtering with Higher Moments (*DsFlsminv1d* and *DsFlsmvsk2d*)

As discussed earlier, many of the despeckle filters proposed in the literature suffer from smoothing effects in edge areas. Because of their statistical working principle, the edges may be better detected by incorporating higher statistical variance moments (variance, skewness, kurtosis) [21] calculated from the local moving window. The variance in every window, i.e., ${}^2_{w}$, may, thus, be described as a function of the variance ^{2}, the skewness ^{3}, and the kurtosis ^{4} in the sliding moving local window and is calculated for the filter *DsFlsminv1d* as (see also Figure 2.1b)

$${}^2_{w} = (c_2 \;\; {}^2 + c_3 \;\; {}^3 + c_4 \;\; {}^4)/(c_2 + c_3 + c_4). \tag{2.6}$$

The constants c_2, c_3, and c_4 in Eq. (2.6) may be calculated using [7]

$$R = 1 \quad \frac{1}{1 + \;\; {}^2}, \tag{2.7}$$

which represents the smoothness of the image. Specifically, the constants c_2, c_3, and c_4 are calculated by replacing the variance ^{2} in Eq. (2.7), the skewness ^{3}, and the kurtosis ^{4} in the moving pixel window, respectively. The higher moments are each weighted with a factor, i.e., c_2, c_3, and c_4, which receives values $0 < c < 1$. Equations (2.6) and (2.7) will be applied in windows where

$$c_3 \;\; {}^3 \quad c_2 \;\; {}^2 \quad c_4 \;\; {}^4. \tag{2.8}$$

In regions where Eq. (2.8) is not valid, the window variance can be calculated as

$${}^2_{w} = (c_2 \;\; {}^2 + c_4 \;\; {}^4)/(c_2 + c_4). \tag{2.9}$$

The final value for ${}^2_{w}$ will be used to replace the variance ^{2} and will be further used for calculating the coefficient of variation in Eq. (2.4). The *DsFlsminv1d* despeckle filter operates in the 1D direction by calculating ${}^2_{w}$ for each row and each column in the sliding moving window (see Figure 2.1b), where the introduction of the higher moments in the filtering process should preserve the edges and should not smooth the image in areas with strong pixel variations. The middle pixel in the window is then replaced with Eq. (2.1) by replacing the $k_{i,j}$ weighting factor with ${}^2_{w}$. ${}^2_{w}$ in Eq. (2.6) can be interpreted as a generalized moment weighting factor with the weighting coefficients c_2, c_3, and c_4. The moving window size for the *DsFlsminv1d* filter was [5 × 5], and its operation is shown in Figure 2.1b.

The despeckle filter *DsFlsmvsk2d* [7] is the 2D realization of *DsFlsminv1d* utilizing the higher statistical moments ^{3} and ^{4} of the image in a [5 × 5] pixel moving window.

Algorithm 2.3 presents the algorithmic steps for the implementation of the *DsFlsmvsk2d* despeckle filter.

ALGORITHM 2.3

	Linear filtering: linear scaling filter (*DsFlsmvsk2d*)
1	Load the image for filtering.
2	Specify the region of interest to be filtered, the moving window size, and the number of iterations (n).
3	Compute the noise variance $^{2}_{n}$ with Eq. (2.5) for the whole image.
4	Starting from the left upper corner of the image, compute for each moving window the coefficient win_var as follows.
5	If Eq. (2.8) is true, use Eq. (2.7); otherwise, use Eq. (2.9).
6	Compute $f_{i,j}$ in Eq. (2.1) and replace the noisy middle point in each moving window $g_{i,j}$ with the new computed value $f_{i,j}$ by using win_var for the coefficient of variation $k_{i,j}$.
7	Repeat steps 4–6 for all the pixels in the image by sliding the moving window from left to right.
8	Repeat steps 3–7 for n iterations.
9	Compute the image quality evaluation metrics and the texture features for the original and despeckled images.
10	Display the original and despeckled images, the image quality and evaluation metrics, and the texture features.

2.1.3 Homogeneous Mask Area Filtering (*DsFlsminsc*)

The *DsFlsminsc* is a 2D filter operating in a [5 × 5] pixel neighborhood by searching for the most homogeneous neighborhood area around each pixel using a [3 × 3] subset window [32], as shown in Figure 2.1c. The middle pixel of the [5 × 5] neighborhood is substituted with the average gray level of the 3 × 3 mask with the smallest speckle index C, which, for log-compressed images, is given by

$$C = {}^{2}_{s} / \bar{g}_s, \tag{2.10}$$

where $^{2}_{s}$ and $\bar{g}_s$ represent the variance and the mean of the [3 × 3] window, respectively. The window with the smallest C is the most homogeneous semiwindow, which presumably does not contain any edge. The filter is iteratively applied until the gray levels of almost all pixels in the image do not change. The operation of the *DsFlsminsc* filter may be described as follows (see also Figure 2.1c).

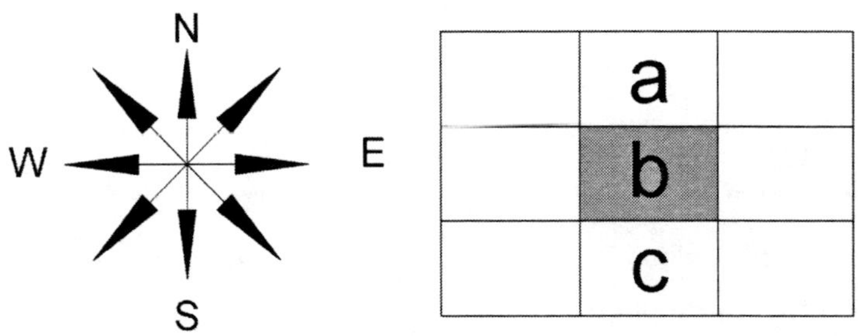

(a) (b)

FIGURE 2.2: (a) Directions of implementation of the *DsFgf4d* geometric filter. (b) Pixels selected for the NS direction (the intensity of central pixel b is adjusted based on the values of intensities of pixels a and b for the geometric filter *DsFgf4d*).

1. Slide the [3 × 3] mask with the [(5 × 5] pixel) selected window.
2. Detect the position of the mask for which C [see Eq. (2.10)] is minimum.
3. Assign the average gray level of the mask to the middle pixel of the [5 × 5] window.
4. Apply steps 1–3 for all pixels in the image.
5. Iterate the above process until the gray levels of almost all pixels in the image do not change.

ALGORITHM 2.4

	Linear filtering: homogeneous mask area filtering (*DsFlsminsc*)
1	Load the image for filtering.
2	Specify the region of interest to be filtered, the moving window size, the number of iterations (n), and the edge detector to be used.
3	Starting from the left upper corner of the image, rotate a mask around the middle pixel of the window for each moving window.
4	Detect the position of the mask for which C [see Eq. (2.10)] is minimum.
5	Assign the average gray level of the mask at the selected position to the middle pixel in the 5 × 5 window.
6	Repeat steps 4 and 5 for all the pixels in the image by sliding the moving window from left to right.
7	Repeat steps 3–6 for a second iteration of despeckle filtering.
8	Compute the image quality evaluation metrics and the texture features for the original and despeckled images.
9	Display the original and despeckled images, the image quality and evaluation metrics, and the texture features.

CODE 2.2

MATLAB code linear filtering: homogeneous mask area filtering (*DsFlsminsc*)

```
1 function f = DsFlsminsc(g, nhood, niterations, edge)
%*****************************************************************************
% Despeckle filtering toolbox, % © Christos P. Loizou 2007
% Ultrasound image-Multiplicative noise filtering
% The filter utilizes different filter detectors, from which you may choose one according to your
% application
% Input variables:
% g :           input image to be filtered, i.e. 'cell.tif'
% nhood :       sliding moving window, i.e [5 5]
% niterations:  iterations for which filtering is applied iteratively
% edge   :edge detector, used for finding the most homogeneous areas within the sliding
window
%        : edge=0, use the variance as an edge detector
%        : edge=1, use the speckle contrast as an edge detector
%        : edge=2, use max|m1-m2| input 2, max|m1/m2, m2/m1|'as an edge detector
%        : edge=3, use the third moment as an edge detector
%        : edge=4, use the fourth moment as an edge detector
%
% Output variable:
2 % f        :input image for filtering
%*****************************************************************************
% Specify the area of interest to be filtered, the moving window size, the number of times
% (niterations) the filtering is applied to the image and the edge detector to be used
disp('Input the edge detector you would like to be used for the filter..');
disp('Input 0 for using the variance as a detector');
disp('Speckle Contrast input 1, max|m1-m2| input 2, max|m1/m2, m2/m1|');
disp('Moment 3rd grades 3, Moment 4th grades 4');
if isa(g, 'uint8')
  u8out = 1;
  if (islogical(g))
     logicalOut = 1;
   g = double(g);
```

```
  else
   logicalOut = 0;
   g = double(g)/255;
  end
else
  u8out = 0;
end
 % Estimate the size of the image
[ma ,na] = size(g);
% Estimate the midle of the processing window, which takes onle values 3, 5, 7
z=(nhood(1)-1)/2;
% Initialize the picture f (new picture) with zeros
f=g;
% Apply the filter niterations on the original image
  for i = 1:niterations
3 fprintf('\rIteration %d',i);
  if i >=2
     g=f;
  end
% Starting from the left upper corner of the image, rotate a mask around the middle pixel of
the % window for each moving window
% Estimate and change the middle pixel in the sliding window
handle=waitbar(0, 'Calculating/replacing the center pixel in a sliding window...');
ini=z+1;
for i= ini :(ma-z)
  for j= ini:(na-z)
    var_neu=1000000000.0;si_neu=10000000000.0; xmit1=0.0; cd_neu=10000000.0;
    hos_neu = 1000000000.0; hos4_neu = 100000000.0;
    for a= (i-z):i
      for b=(j-z):j
        xmit= 0.0;
        for l=a:(a+z)
          for p=b:(b+z)
            xmit=xmit + g(l, p);
        end
```

```
4 % End for p
        end
 % End for l
        xmit = (1.0/9.0) *xmit;
        var=0.0; pk=0.0; pk4=0.0;
% Detect the position of the mask for which the C of the gray levels is minimum
        for l=a:(a+z)
          for p=b:(b+z)
            var= var + ((g(l,p)-xmit)*(g(l, p)-xmit) );
           % 3rd moment
            pk= pk + ((g(l,p)-xmit)*(g(l, p)-xmit) * (g(l, p)-xmit) );
            pk4=pk4 +(g(l,p)-xmit)*(g(l, p)-xmit)*(g(l, p)-xmit)*(g(l,p)-xmit);
          end
% End for p
        end
5 % End for l
% Variance in subwindow
        var = (1/9.0)* var;
% 3rd moment in window
        pk = (1/9.0)*pk;
 % Assign the average gray level of the mask at the selected position to the middle pixel
      if xmit ~=0.0
% Speckle index in subwindow
          si = sqrt(var)/xmit;
        else
          si=0.0;
        end
% Gradient information of the subset
        cd = abs(xmit-xmit1);
        xmit1 = xmit;
        if xmit~=0.0
% 3rd higher order statistics
          hos = power(pk, 0.5) /xmit;
        % 4th higher order statistics
          hos4 = power (pk4, 0.25)/xmit;
        else
```

```
      hos=0.0;
      hos4=0.0;
    end
    % Use the speckle contrast to calculate f(i, j)
    if edge == 1
      if si < si_neu
      si_neu = si;
      f(i, j) = xmit;
      end
    % end if speckle index
    elseif edge == 0
% Use the variance to calculate f(i, j)
      if var <var_neu
      var_neu = var;
      f(i, j) = xmit;
      end
     % end if var
    elseif edge == 2
      if cd < cd_neu
        cd_neu =cd;
% Use the local gradient to calculate f(i, j)
        f(i, j)= xmit;
      end
% End if local gradient
    elseif edge == 3
      if hos < hos_neu
          hos_neu = hos;
% Use higher moments to calculate f(i, j)
        f(i, j) = xmit;
      end
% end if higher order statistics 3rd grades
    elseif edge == 4
      if hos4 < hos4_neu
        hos4_neu = hos4;
% use higher moments to 4rth grades to calculate f(i, j)
        f(i, j) = xmit;
```

```
            end                % end if higher order statistics 4rth grades
          end                  % end if edge

        end   % end for b
      end     % end for a
6 % Repeat steps 4 and 5 for all the pixels in the image by sliding the moving window from
  left to
  % right
    end     %end for n
       waitbar(i/na)
  end       % end for m
7
   % Repeat steps 3 to 6 for a second iteration of despeckle filtering
  close(handle)
  end
  % End for itterations
  fprintf('\n');

  if u8out==1,
8  if (logicalOut)
     f = uint8(f);
   else
     f = uint8(round(f*255));
   end
  end
  % Calculate 61 Texture Features from the original and despeckled images
  A=[ ]; F1=[ ] ;
  %  Initialize the matrcies for texture features
  T= DsTtexfeat(double(g));
  A=[A, T'];
  % Save the texture features of the original image in a matrix A
  save or_texfeats A;

  TAM=DsTtexfeat(double(f));
9 F1=[F1,TAM'];
  save speckle1texfs F1;
  % The texture features of the despeckled image are saved in matrix F1
```

```
% Call the function metrics to calculate and display the 19 different image quality metrics
between % the original and the despeckled image
M=DsQmetrics(f, g);

% Display both the original and the despeckled images on the screen
figure, imshow(g), title('Original Image');
figure, imshow(f), title('Image filtered by maskedge filter');
```

Algorithm 2.4 presents the algorithmic steps for the implementation of the *DsFlsminsc* despeckle filter, whereas Code 2.2 presents the implementation of the filter in MATLAB™ code.

Figure 4.1c shows the application of the despeckle filter *DsFlsminsc* on a phantom carotid artery image for a moving window size of [5 × 5] pixels and three iterations. Figure 4.4c shows the results of the filter on an artificial carotid image, whereas Figure 4.6c presents the results on a real carotid ultrasound image.

2.2 NONLINEAR FILTERING

Nonlinear filtering is based on nonlinear operations involving the pixels in a neighborhood. For example, letting the center pixel in the moving window be equal to the maximum pixel in its neighborhood is a nonlinear filtering operation.

2.2.1 Median Filtering (*DsFmedian*)

The filter *DsFmedian* [33] is a simple nonlinear operator that replaces the middle pixel in the window with the median value of its neighbors. The moving window for the *DsFmedian* filter used for the experiments presented in this book was [7 × 7].

Algorithm 2.5 presents the algorithmic steps for the implementation of the *DsFmedian* despeckle filter.

ALGORITHM 2.5

Nonlinear filtering: median filter (*DsFmedian*)

1 Load the image for filtering.

2 Specify the region of interest to be filtered, the moving window size, and the number of iterations (n).

3 Starting from the left upper corner of the image, compute for each sliding moving window its median value.

4	Replace the middle pixel in the sliding window with the median value calculated in step 3.
5	Repeat steps 3 and 4 for the whole image by sliding the moving window from left to right.
6	Repeat steps 3–5 for *n* iterations.
7	Compute the image quality evaluation metrics and the texture features for the original and despeckled images.
8	Display the original and despeckled images, the image quality and evaluation metrics, and the texture features.

Figure 4.1e shows the application of the despeckle filter *DsFmedian* on a phantom carotid artery image for a moving window size of [5 × 5] pixels and four iterations. Figure 4.4e shows the results of the filter on an artificial carotid image, whereas Figure 4.6e presents the results on a real carotid ultrasound image.

2.2.2 Linear Scaling Filter (*DsFca*, *DsFlecasort*, and *DsFls*)

The DsFca filter despeckles the image through linear scaling of the gray-level values [46]. In a window of [5 × 5] pixels, compute the mean of all pixels whose difference in the gray level with the intensity $g_{i,j}$ (the middle pixel in the moving window) is lower than or equal to a given threshold . Assign this value to the gray level $g_{i,j}$ with = g_{max}, where g_{max} is the maximum gray level of the image and = [0,1]. Best results can be obtained with = 0,1.

The *DsFlecasort* filter [46] takes *k* points of a pixel neighborhood, which are closest to the gray level of the image at point $g_{i,j}$ (the middle point in the moving window), including $g_{i,j}$ [61]. It then assigns the mean value of these points to the pixel $g_{i,j}$ (usually, $N = 9$ in a 3 × 3 window, where $k = 6$).

The *DsFls* filter [46] scales the pixel intensities by finding the maximum g_{max} and the minimum g_{min} gray-level values in every moving window, and then replaces the middle pixel with

$$f_{i,j} = \frac{g_{max} + g_{min}}{2}. \tag{2.11}$$

Algorithm 2.6 presents the algorithmic steps for the implementation of the *DsFlsminsk2d* despeckle filter.

ALGORITHM 2.6

	Nonlinear filtering: linear scaling filter (*DsFca*)
1	Load the image for filtering.
2	Specify the region of interest to be filtered, the moving window size, and the number of iterations (*n*).

3 Starting from the left upper corner of the image, compute for each moving window the mean of all pixels in the window whose difference in the gray level with the middle pixel in the moving window is lower than or equal to a given threshold , with = $g_{\max}$.

4 Assign the computed mean value to the middle pixel in the window.

5 Repeat steps 3 and 4 for *n* iterations.

6 Compute the image quality evaluation metrics and the texture features for the original and despeckled images.

7 Display the original and despeckled images, the image quality and evaluation metrics, and the texture features.

Figure 4.1f shows the application of the despeckle filter *DsFls* on a phantom carotid artery image for a moving window size of [5 × 5] pixels and three iterations. Figure 4.4f shows the results of the filter on an artificial carotid image.

2.2.3 Maximum Homogeneity Over Pixel Neighborhood Filtering (*DsFhomog*)

The *DsFhomog* filter is based on an estimation of the most homogeneous neighborhood around each image pixel [34]. The filter takes into consideration only pixels that belong in the processed neighborhood ([7 × 7] pixels) using Eq. (2.12) under the assumption that the observed area is homogeneous. The output image is then given by

$$f_{i,j} = (c_{i,j} g_{i,j}) / \ _{i,j} c_{i,j} \text{ with } c_{i,j} = 1 \text{ if } (1 \quad 2 \quad _n)\bar{g} \quad g_{i,j} \quad (1 + 2 \quad _n)\bar{g} \qquad (2.12)$$

$$c_{i,j} = 0 \text{ otherwise.}$$

The *DsFhomog* filter does not require any parameters or thresholds to be tuned, thus making the filter suitable for automatic implementation.

Algorithm 2.7 presents the algorithmic steps for the implementation of the *DsFhomog* despeckle filter.

ALGORITHM 2.7

Nonlinear filtering: maximum homogeneity filter (*DsFhomog*)

---

1 Load the image for filtering.

2 Specify the region of interest to be filtered, the moving window size, and the number of iterations (*n*).

3	Starting from the left upper corner of the image, apply for each moving window Eq. (2.12) and replace the middle pixel with the new value.
4	Repeat step 3 for the whole image by sliding the moving window from left to right.
5	Repeat steps 3 and 4 for n iterations.
6	Compute the image quality evaluation metrics and the texture features for the original and despeckled images.
7	Display the original and despeckled images, the image quality and evaluation metrics, and the texture features.

Figure 4.1g shows the application of the despeckle filter *DsFhomog* on a phantom carotid artery image for a moving window size of [7 × 7] pixels and five iterations. Figure 4.4g shows the results of the filter on an artificial carotid image, whereas Figure 4.6f presents the results on a real carotid ultrasound image.

2.2.4 Geometric Filtering (*DsFgf4d*)

The concept of geometric filtering is that speckle appears in the image as narrow walls and valleys. The geometric filter, through iterative repetition, gradually tears down the narrow walls (bright edges) and fills up the narrow valleys (dark edges), thus smearing the weak edges that need to be preserved.

The *DsFgf4d* filter [10] uses a nonlinear noise reduction technique. It compares the intensity of the central pixel in a [3 × 3] neighborhood with those of its eight neighbors, and, based on the neighborhood pixel intensities, it increments or decrements the intensity of the central pixel such that it becomes more representative of its surroundings. The operation of the geometric filter *DsFgf4d* may be described with Figure 2.2a and 2.2b and has the following form.

1. Select the direction and assign the pixel values.
   Select the direction to be north–south (NS) and the corresponding three consecutive pixels to be a, b, and c (see Figure 2.2a and 2.2b, respectively).
2. Carry out central pixel adjustments.
   Do the following intensity adjustments (see Figure 2.2b):

if $a \quad b+2$, then $b=b+1$;
if $a \succ b$ and $b \quad c$, then $b=b+1$;
if $c \succ b$ and $b \quad a$, then $b=b+1$;
if $c \quad b+2$, then $b=b+1$;
if $a \quad b \quad 2$, then $b=b \quad 1$;
if $a \prec b$ and $b \quad c$, then $b=b \quad 1$;
if $c \prec b$ and $b \quad a$, then $b=b \quad 1$;
if $c \quad b \quad 2$, then $b=b \quad 1$.

3. Repeat.

   Repeat steps 1 and 2 for directions west–east (WE), west–north to southeast (WN–SE), and northeast to west–south (NE–WS; see Figure 2.2a).

ALGORITHM 2.8

Nonlinear filtering: geometric filter (*DsFgf4d*)	
1	Load the image for filtering.
2	Specify the region of interest to be filtered, the moving window size, and the number of iterations (n) the filtering is applied to the image.
3	Starting from the left upper corner of the image, within the [5 × 5] pixel moving window, rotate a 3 × 3 pixel mask around the middle pixel of the window.
4	Carry out pixel adjustments as explained above.
5	Assign the new grayscale value of the selected position to the middle pixel.
6	Repeat steps 4 and 5 for all the pixels in the image by sliding the moving window from left to right.
7	Repeat steps 3–6 for n iterations.
8	Compute the image quality evaluation metrics and the texture features for the original and despeckled images.
9	Display the original and despeckled images, the image quality and evaluation metrics, and the texture features.

CODE 2.3

MATLAB code nonlinear filtering: geometric filter (*DsFgf4d*)

```
function f = DsFgf4d(g, nhood, niterations)
%*******************************************************************
% Despeckle filtering toolbox, © Christos P. Loizou 2007
% Speckle reduction filter: gf4d
% A non-linear geometric filter that filters the multiplicative noise in ultrasound Images.
Utilizes % the local statistics of the noise (original) image g
% Input Variables:
% g :          input  image  to be filtered, i.e. 'cell.tif'
% nhood :     sliding moving window, i.e [5 5]
```

```
% niterations:   itterations for which filtering is applied iteratively
%
% Output Variables:
% f: outpout image
% Call:  f = DsFgf4d (g, [5 5], 4);
%*****************************************************************
if isa(g, 'uint8')
  u8out = 1;
  if (islogical(g))
    logicalOut = 1;
    g = double(g);
  else
    logicalOut = 0;
    g = double(g)/255;
  end
else
  u8out = 0;
end

% Specify the area of interest to be filtered, the moving window size and the number of
2 itearions
% (n) the filtering is applied to the image

% Estimate the size of the image
[ma ,na] = size(g);
% Estimate the midle of the processing window, which takes onle values 3, 5 7
z=(nhood(1)-1)/2;
%Initialize the picture f (new picture) with zeros
f=g;


3 % Starting from the left upper corner of the image, rotate a 5x5 pixel mask around the middle
pixel % of the window for each moving window
for i = 1:niterations
  fprintf('\rIteration %d',i);
  if i >=2
     g=f;
  end
```

```
4 % Carry out pixel adjustments as explained above (Estimate and change the middle pixel in
  % the window
  disp(['      Calculating/replacing the center pixel in a sliding window...']);
  %ma=100; na=100;
  a=1; b=0; c=3; d=1;
  while c>=0,
    for d=0:1
      for i= 2 :(ma-1)
        for j= 2:(na-1)
          maxi= min(g(i-a, j-b)-1, g(i, j) +1);
          f(i, j) = max(g(i, j), maxi);
        end
      % End for j
      end
      % End for i
     for i= 2 :(ma-1)
        for j= 2:(na-1)
          maxin1 = min(f(i-a, j-b), g(i, j) +1);
          maxin = min(maxin1, f(i+a, j+b)+1);
  % Assign the new greyscale value at the selected position to the middle pixel
5         g(i, j) = max(f(i, j),  maxin);
        end
  % End for j
     end
  % End for i
     if d==0
        a=-a; b=-b;
      end


    end
  % End if d
    disp(['First Itteration of the Algorithm is Applied']);

    for d=0:1
      for i= 2 :(ma-1)
```

```
          for j= 2:(na-1)
            mini = max(g(i-a, j-b)+1, g(i, j) -1);
            f(i, j) = min(g(i, j), mini );
5         end
  % End for j
        end
   % End for i
      for i= 2 :(ma-1)
          for j= 2:(na-1)
            mini1 = max(f(i-a, j-b), g(i, j) -1);
            minin = max (mini1, f(i+a, j+b)-1);
            g(i, j) = min(f(i, j),  minin);
          end
   % End for j
5     end
      % End for i
      if d==0
         a=-a; b=-b;
       end
  % End if d
    end
  % End for d
    disp(['Second Itteration of the Algorithm is Applied']);
  % Repeat steps 4 and 5 for all the pixels in the image by sliding the moving window from left
  to
  % right
6  switch c
  case 3
    a=0; b=1; c=2;
    break;
  case 2
    a=1; b=1; c=1;
    break;
  case 1
    a=1; b=-1; c=0;
    break;
  case 0
```

```
    c=-1;
    break;
  end
  % End switch
  end
  % End while c>=0 loop
  end
  % Repeat steps 3 to 6 for n iterations
  fprintf('\n');

7 if u8out==1,
    if (logicalOut)
      f = uint8(f);
    else
      f = uint8(round(f*255));
    end
  end
   figure, imshow(f);
   title('Image filtered by gf4d filter');
  % Compute the texture features and image quality evaluation metrics and display both the
  original % and the despeckled images on the screen
  TAM=DsTtexfeat(double(f));
  F1=[F1,TAM'];
  save speckle1texfs F1;
8 % The texture features of the despeckled image are saved in matrix F1
  % Call the function metrics to calculate and display the 19 different image quality metrics
  between % the original and the despeckled image
  M=DsQmetrics(f, imagec);

  % Show original and despeckled images on the screen
  figure, imshow(a), title ('Original Image');
  figure, imshow(f), title ('Despeckled Image');
```

Algorithm 2.8 presents the algorithmic steps for the implementation of the *DsFgf4d* despeckle filter together with the MATLAB™ code.

Figure 4.1h shows the application of the despeckle filter *DsFgf4d* on a phantom carotid artery image for a moving window size of [7 × 7] pixels and four iterations. Figure 4.4h shows the results of the filter on an artificial carotid image, whereas Figure 4.6g presents the results on a real carotid ultrasound image. Finally, in Figure 4.7d and 4.7h, examples of the filter are shown applied on cardiac ultrasound images.

2.2.5 Homomorphic Filtering (*DsFhomo*)

The *DsFhomo* filter performs homomorphic filtering for image enhancement by computing the fast Fourier transform (FFT) of the logarithmic compressed image, applying a denoising homomorphic filter function $H(\cdot)$, and then performing the inverse FFT of the image [17, 18]. The homomorphic filter function $H(\cdot)$ may be constructed using either a band-pass Butterworth or a high-boost Butterworth filter. In this book, a high-boost Butterworth filter was used with the homomorphic function [17]

$$H_{u,v} = {}_{\mathrm{L}} + \frac{{}_{\mathrm{H}}}{1 + (D_0/D_{u,v})^2} \tag{2.13a}$$

with

$$D_{u,v} = \overline{(u \quad N/2)^2 + (\quad N/2)^2}, \tag{2.13b}$$

where $D_0 = 1.8$ is the cut of the frequency of the filter, ${}_{\mathrm{L}} = 0.4$ and ${}_{\mathrm{H}} = 0.6$ are the gains for the low and high frequencies, respectively, u and   are the spatial coordinates of the frequency transformed image, and N is the dimension of the image in the u,   space.

This form of filtering sharpens the features and flattens the speckle variations in an image.

Algorithm 2.9 presents the algorithmic steps for the implementation of the *DsFhomo* despeckle filter.

Figure 4.1i shows the application of the despeckle filter *DsFhomo* on a phantom carotid artery image for a moving window size of [5 × 5] pixels and four iterations. Figure 4.4i shows the results of the filter on an artificial carotid image, whereas Figure 4.6h presents the results on a real carotid ultrasound image.

ALGORITHM 2.9

	Nonlinear filtering: homomorphic filter (*DsFhomo*)
1	Load the image for filtering.
2	Calculate the FFT of the image.
3	Construct a denoising homomorphic filter using H as in Eq. (2.13).

4 Apply the denoising homomorphic filter function H calculated in step 3.

5 Perform the inverse FFT of the image to form the despeckled image.

6 Compute the image quality evaluation metrics and the texture features for the original and despeckled images.

7 Display the original and despeckled images, the image quality and evaluation metrics, and the texture features.

2.3 DIFFUSION FILTERING

Diffusion filters remove the noise from an image by modifying the image via solving a partial differential equation (PDE). Smoothing is carried out depending on the image edges and their directions. Anisotropic diffusion is an efficient nonlinear technique for simultaneously performing contrast enhancement and noise reduction. It smooths homogeneous image regions, but retains image edges [5, 22, 23] without requiring any information from the image power spectrum. It may, thus, directly be applied to images.

Consider applying the isotropic diffusion equation given by $\mathrm{d}g_{i,j,t}/\mathrm{d}t = \mathrm{div}(d\ \ g)$ using the original noisy image $g_{i,j,t=0}$ as the initial condition, where $g_{i,j,t=0}$ is an image in the continuous domain, i and j specify the spatial position, t is an artificial time parameter, d is the diffusion constant, and $\ g$ is the image gradient. Modifying the image according to this linear isotropic diffusion equation is equivalent to filtering the image with a Gaussian filter. In this chapter, we will present conventional anisotropic diffusion (*DsFad*), speckle-reducing anisotropic diffusion (*DsFsrad*), and coherent nonlinear anisotropic diffusion (*DsFnldif*).

2.3.1 Anisotropic Diffusion Filtering (*DsFad*)

Perona and Malik [23] replaced the classical isotropic diffusion equation, as described above, by the introduction of a function $d_{i,j,t} = f(|\ \ g|)$ that smooths the original image while trying to preserve brightness discontinuities with

$$\frac{\mathrm{d}g_{i,j,t}}{\mathrm{d}t} = \mathrm{div}[d_{i,j,t}\ \ g_{i,j,t}] = \left[\frac{\mathrm{d}}{\mathrm{d}i} d_{i,j,t} \frac{\mathrm{d}}{\mathrm{d}i} g_{i,j,t}\right] + \left[\frac{\mathrm{d}}{\mathrm{d}j} d_{i,j,t} \frac{\mathrm{d}}{\mathrm{d}j} g_{i,j,t}\right], \tag{2.14a}$$

where $|\ \ g|$ is the gradient magnitude, and $d(|\ \ g|)$ is an edge stopping function, which is chosen to satisfy $d\ \ \ 0$ when $|\ \ g|\ \ \ \ $, so that the diffusion is stopped across edges. This function, called the diffusion coefficient $d(|\ \ g|)$, which is a monotonically decreasing function of the gradient magnitude $|\ \ g|$, yields intraregion smoothing and not interregion smoothing [19, 20, 22, 23] by impeding the diffusion at image edges. It increases smoothing parallel to the edge and stops smoothing perpendicular to the edge, as the highest gradient values are perpendicular to the edge and dilated across edges. The choice of $d(|\ g|)$ can greatly affect the extent to which discontinuities are preserved. For example, if $d(|\ g|)$ is constant at all locations, then smoothing progresses in an isotropic manner. If $d(|\ g|)$ is allowed to vary according to the local image gradient, then we have anisotropic diffusion. A basic anisotropic PDE is given in Eq. (2.14a). Two different diffusion coefficients were proposed in Ref. [23] and also derived in Ref. [22]. The diffusion coefficients suggested were

$$d(|\ g|) = \frac{1}{1 + (|\ g_{i,j}|/K)^2} \text{ and } cd(|\ g|) = \frac{2(|\ g_{i,j}|)}{2 + (|\ g_{i,j}|/K_1)^2} \tag{2.14b}$$

where K and K_1 are positive gradient threshold parameters, known as diffusion or flow constants [22]. In this book, the first diffusion coefficient in Eq. (2.14b) was used as it was found to perform better in our images [7, 24].

A discrete formulation of the anisotropic diffusion in Eq. (2.14a) is [2, 22, 23]

$$\frac{dg_{i,j}}{dt} = \frac{}{|\ _s|}\{d_{i+1,j,t}[g_{i+1,j}\ \ g_{i,j}] + d_{i\ \ 1,j,t}[g_{i\ \ 1,j}\ \ g_{i,j}] + d_{i,j+1,t}[g_{i,j+1}\ \ g_{i,j}] + d_{i,j\ \ 1,t}[g_{i,j\ \ 1}\ \ g_{i,j}]\} \tag{2.15a}$$

where the new pixel gray value $f_{i,j}$ at location i,j is

$$f_{i,j} = g_{i,j} + \frac{1}{4}\frac{dg_{i,j}}{dt}, \tag{2.15b}$$

where $d_{i+1,j,t}$, $d_{i\ \ 1,j,t}$, $d_{i,j+1,t}$, and $d_{i,j\ \ 1,t}$ are the diffusion coefficients for the west, east, north, and south pixel directions, respectively, in a four-pixel neighborhood around the pixel i,j where diffusion is computed. The diffusion coefficient leads to the largest diffusion where the nearest-neighbor difference is largest (the largest edge), whereas the smallest diffusion is calculated where the nearest-neighbor difference is smallest (the weakest edge). The constant $\ \ \Re^+$ is a scalar that determines the rate of diffusion, η_s represents the spatial neighborhood of pixel i,j, and $|\eta_s|$ is the number of neighbors (usually four except at the image boundaries). Perona and Malik [23] linearly approximated the directional derivative in a particular direction as $\ g_{i,j} = g_{i+1,j}\ \ g_{i,j}$ (for the east direction of the central pixel i,j). Modifying the image according to Eq. (2.15), which is a linear isotropic diffusion equation, is equivalent to filtering the image with a Gaussian filter. The parameters for the anisotropic diffusion filter used in this book were $\ = 0.25$, $\ _s = 8$, and $K = 30$, which was used for the calculation of the edge stopping function $d(|\ g|)$, in Eq. (2.14b).

Figure 4.4j shows the results of the filter *DsFad* on an artificial carotid image for 20 iterations, whereas Figure 4.6i presents the results on a real carotid ultrasound image.

2.3.2 Speckle-Reducing Anisotropic Diffusion Filtering (*DsFsrad*)

The essence of speckle-reducing anisotropic diffusion is the replacement of the gradient-based edge detector $cd(|\ g|)$ in an original anisotropic diffusion PDE with the instantaneous coefficient of variation that is suitable for speckle filtering $c_{srad}(| \ g|)$. The *DsFsrad* speckle-reducing anisotropic diffusion filter [5] uses two seemingly different methods, namely, the Lee [9, 14, 15] and Frost diffusion filters [13]. A more general updated function for the output image by extending the PDE versions of the despeckle filter is [5]

$$f_{i,j} = g_{i,j} + \frac{1}{s} \mathrm{div}(c_{srad}(| \ g|) \ g_{i,j}). \tag{2.16}$$

The diffusion coefficient for the speckle anisotropic diffusion $c_{srad}(| \ g|)$ is derived [5] as

$$c^2_{srad}(| \ g|) = \frac{\frac{1}{2}| \ g_{i,j}|^2 \ \frac{1}{16}(\ ^2 g_{i,j})^2}{(g_{i,j} + \frac{1}{4} \ ^2 g_{i,j})^2}. \tag{2.17}$$

It is required that $c_{srad}(| \ g|) \ 0$. The above instantaneous coefficient of variation combines a normalized gradient magnitude operator and a normalized Laplacian operator to act like an edge detector for speckle images. A high relative gradient magnitude and a low relative Laplacian indicate an edge. The *DsFsrad* filter utilizes speckle-reducing anisotropic diffusion after Eq. (2.15) with the diffusion coefficient $c_{srad}(| \ g|)$ in Eq. (2.17) [5].

ALGORITHM 2.10

Diffusion filtering: speckle-reducing anisotropic filter (*DsFsrad*)

1 Load the image for filtering.

2 Specify the original image to be filtered (K), the number of iterations (n), the time step (lambda), and the region of interest to be filtered (rect).

3 Transform the original image to double and normalize it to $f_{i,j} = (f_{i,j} \ \text{min pixelvalue})/(\text{max pixelvalue} \ \text{min pixelvalue})$, where min pixelvalue and max pixelvalue represent the minimum and maximum pixel values in the image.

4 Starting from the left upper corner of the image, select a [3 × 3] pixel neighborhood and compute a new grayscale value according to Eq. (2.16).

5 Assign the new grayscale value to the middle pixel in each window.

6 Repeat steps 4 and 5 for the whole image by sliding the moving window from left to right.

7 Repeat steps 4–6 for niter iterations.

8 Compute the image quality evaluation metrics and the texture features for the original and despeckled images.

9 Display the original and despeckled images, the image quality and evaluation metrics, and the texture features.

CODE 2.4

MATLAB code diffusion filtering: speckle-reducing anisotropic filter (*DsFsrad*)

```
1 function [I,rect] = DsFSRAD(K,niter,lambda,rect)
  %*************************************************************************
2 % Load the image for filtering
  % Specify the area of interest to be filtered (I), the number of iterations (niter), the time step
  % (lamda), and the area to be filtered (rect)
  % Despeckle filtering toolbox, % © Christos P. Loizou 2007
  % Speckle filtering using SRAD (Speckle Reducing Anisotropic Diffusion)
  % Input Variables:
  % K: original image
  % niter = number of iterations to apply the filter
  % lambda = time step
  % rect: rectangle area to be filtered
  %
  % Output Variables:
  % I = new smoothed image
  % rect = region of interest (ROI)
  %
  % Example 1: [I,rect] = DsFSRAD(K(:,:,1),75,0.025);
  % Example 2: [I,rect] = DsFSRAD(K(:,:,1),75,0.025, [0 0 436 182]);
  % to despeckle the image by directly defining the ROI to be filtered
3 %*************************************************************************
  % Transform the original image to double and normalize it
  I = double(K);
  mx = max(I(:));
  mn = min(I(:));
  I = (I-mn)/(mx-mn);
  % indices (using boudary conditions)
  [M,N] = size(I);
  iN = [1, 1:M-1];
  iS = [2:M, M];
  jW = [1, 1:N-1];
  jE = [2:N, N];
  % get an area of uniform speckle
```

```
if nargin < 4 || isempty(rect)
   imshow(I,[],'notruesize');
   rect = getrect;
end
% log uncompress the image and eliminate zero value pixels.
4 I = exp(I);
% wait bar
hwait = waitbar(0,'Diffusing Image');
% Starting from the left upper corner of the image, select a 3x3 pixel neighbourhood and
compute a % new greyscale value according to (2.16)
for iter = 1:niter
    % speckle scale function
    Iuniform = imcrop(I,rect);
    q0_squared = (std(Iuniform(:))/mean(Iuniform(:)))^2;

    % differences
    dN = I(iN,:) - I;
    dS = I(iS,:) - I;
    dW = I(:,jW) - I;
    dE = I(:,jE) - I;
     % normalized discrete gradient magnitude squared
    G2 = (dN.^2 + dS.^2 + dW.^2 + dE.^2) ./ (I.^2 + eps);
    % normalized discrete laplacian
    L = (dN + dS + dW + dE) ./ (I + eps);
     % ICOV (equ 31/35)
    num = (.5*G2) - ((1/16)*(L.^2));
    den = (1 + ((1/4)*L)).^2;
    q_squared = num ./ (den + eps);
     % diffusion coefficent
    den = (q_squared - q0_squared) ./ (q0_squared *(1 + q0_squared) + eps);
    c = 1 ./ (1 + den);
    cS = c(iS, :);
    cE = c(:,jE);
5    % divergence
    D = (cS.*dS) + (c.*dN) + (cE.*dE) + (c.*dW);
6    % Assign the new greyscale value to the middle pixel in each window
```

```
      I = I + (lambda/4)*D;
   % Repeat steps 4 and 5 for the whole image by sliding the moving window from left to right
      waitbar(iter/niter,hwait);
7  end
   % Repeat steps 4 to 6 for niter iterations
   I = log(I);
   figure, imshow(I);
   % close wait bar
   close(hwait)
8  return;
   % Compute the texture features and image quality evaluation metrics and display both the
   original % and the despeckled image on the screen
   TAM=DsTtexfeat(double(I));
   F1=[F1,TAM'];
   save speckle1texfs F1;
   % The texture features of the despeckled image are saved in matrix F1
   % Call the function metrics to calculate and display the 19 different image quality metrics
   between %  the original and the despeckled image
   M=DsQmetrics(I, K);

   % Show original and despeckled images on the screen
   figure, imshow(K), title ('Original Image');
   figure, imshow(I), title ('Despeckled Image');
```

Algorithm 2.10 presents the algorithmic steps for the implementation of the *DsFsrad* filter together with the MATLAB™ code.

Figure 4.7f shows the results of the *DsFsrad* filter on a cardiac ultrasound image, with a coefficient of variation of 0.025 and 50 iterations.

2.3.3 Coherent Nonlinear Anisotropic Diffusion Filtering (*DsFnldif*)

The applicability of the *DsFad* filter [see Eq. (2.16)] is restricted to smoothing with edge enhancement, where $| \ g|$ has higher magnitude at edges. In general, the function $d(| \ g|)$ in Eq. (2.16) can be put into a tensor form that measures the local coherence of structures such that the diffusion process becomes more directional in both the gradient and contour directions, which represent the directions of maximum and minimum variations, respectively. Therefore, the *DsFnldif* filter will take the form

$$\frac{dg_{i,j,t}}{dt} = \mathrm{div}[D \quad g] \tag{2.18}$$

where $D \quad \Re^{2\times 2}$ is a symmetric positive semidefinite diffusion tensor representing the required diffusion in both gradient and contour directions and, hence, enhancing coherent structures as well as edges. The design of D, as well as the derivation of the coherent nonlinear anisotropic diffusion model, may be found in Ref. [24] and is given as

$$D = (\ _1 \ _2)\begin{pmatrix} \ _1 & 0 \\ 0 & \ _2 \end{pmatrix}\begin{pmatrix} \ _1^T \\ \ _2^T \end{pmatrix} \tag{2.19a}$$

with

$$\ _1 = \begin{cases} \left(1 \quad \frac{(\ _1 \quad \ _2)^2}{s^2}\right) & \text{if} (\ _1 \ _2)^2 \quad s^2 \\ 0 & \text{else} \end{cases} \tag{2.19b}$$

$$\ _2 =$$

where the eigenvectors $\ _1$ and $\ _2$ and the eigenvalues $\ _1$ and $\ _2$ correspond to the directions of maximum and minimum variations and the strength of these variations, respectively. The flow at each point is affected by the local coherence, which is measured by $(\ _1 \quad \ _2)$ in Eq. (2.19b).

The parameters used in this book for the *DsFnldif* filter were $s^2 = 2$ and $\ = 0.9$, which were used for the calculation of the diffusion tensor D and the parameter step size $m = 0.2$, which defined the number of diffusion steps performed. The local coherence is close to zero in very noisy regions, and the diffusion becomes isotropic ($\ _1 = \ _2 = \ = 0.9$), whereas in regions with lower speckle noise, the local coherence corresponds to $(\ _1 \quad \ _2)^2 > s^2$ [24].

Algorithm 2.11 presents the algorithmic steps for the implementation of the *DsFnldif* despeckle filter.

ALGORITHM 2.11

Diffusion filtering: coherent nonlinear anisotropic diffusion filter (*DsFnldif*)	
1	Load the image for filtering.
2	Specify the area of interest to be filtered, the number of iterations (n), and the time step.
3	Transform the image to double and normalize it.
4	Starting from the left upper corner of the image, select a [3 × 3] pixel neighborhood and compute a new grayscale value according to Eqs. (2.18) and (2.19).
5	Assign the new grayscale value to the middle pixel in each window.
6	Repeat steps 4 and 5 for n iterations by sliding the moving window from left to right.

7	Repeat steps 4–6 for the whole image.
8	Compute the image quality evaluation metrics and the texture features for the original and despeckled images.
9	Display the original and despeckled images, the image quality and evaluation metrics, and the texture features.

Figure 4.4k shows the results of the *DsFnldif* filter on an artificial carotid image, whereas Figure 4.6j shows the results of the *DsFnldif* filter on a carotid ultrasound image for 25 iterations.

2.4 WAVELET FILTERING (*DsFwaveltc*)

Wavelet filtering exploits the decomposition of the image into the wavelet basis and zeroes out the wavelet coefficients to despeckle the image [25–29, 38].

Wavelet analysis is particularly useful for the analysis of transient, nonstationary, or time-varying signals. Wavelets can be used to analyze signals in different spatial resolutions. Their advantage is in their ability to analyze a signal with accuracy in both the time and frequency domains. This is not the case when applying traditional Fourier analysis, where there is significant accuracy in the frequency domain, but less accuracy in the temporal domain. In other words, increasing accuracy in one domain implies a decrease in precision in the other domain. Wavelets are also known for their capacity to identify singularities associated with fine variations of the signal to be evaluated [29]. For denoising, we need to identify the specific image scales where most of the image energy lies.

Speckle reduction filtering in the wavelet domain is based on the idea of the Daubechies Symlet wavelet and on soft-thresholding denoising. It was firstly proposed by Donoho [29] and also further investigated by Zhong and Cherkassky [25], Achim et al. [26], and Gupta et al. [38]. The Symlet family of wavelets, although not perfectly symmetrical, was designed to have the least asymmetry and the highest number of vanishing moments for a given compact support [29]. The *DsFwaveltc* filter implemented in this study is described as follows.

1. Estimate the variance of the speckle noise $\ ^2_n$ with Eq. (2.5).
2. Compute the discrete wavelet transform (DWT) using the Symlet wavelet for two scales.
3. For each subband:

   (a) Compute a threshold [27, 29]

$$T = \begin{cases} (T_{max} \quad (j \quad 1)) \ _n & \text{if } T_{max} \quad (j \quad 1) \quad T_{min} \\ T_{min} \ _n & \text{else} \end{cases} \tag{2.20}$$

where is a decreasing factor between two consecutive levels, T_{max} is a maximum factor for $_n$, and T_{min} is a minimum factor for $_n$. The threshold T is primarily calculated using $_n$ and a decreasing factor T_{max} $(j\ \ 1)$.

(b) Apply the threshold on the wavelet coefficients of each band.

4. Compute the inverse DWT to reconstruct the despeckled image *f*.

Algorithm 2.12 presents the algorithmic steps for the implementation of the *DsFwaveltc* despeckle filter.

ALGORITHM 2.12

	Wavelet filtering: wavelet filter (*DsFwaveltc*)
1	Load the image for filtering.
2	Specify the area of interest to be filtered, the moving window size, and the number of iterations (n) the filtering is applied to the image.
3	Compute the noise variance ^2_n with Eq. (2.5) from the whole image.
4	Compute the DWT using the Symlet wavelet for two scales.
5	Compute for each subband a threshold T according to Eq. (2.20).
6	Apply the threshold on the wavelet coefficients for each band.
7	Compute the inverse DWT to reconstruct the whole image.
8	Repeat steps 3–6 for n iterations for the whole image.
9	Compute the image quality evaluation metrics and the texture features for the original and despeckled images.
10	Display the original and despeckled images, the image quality and evaluation metrics, and the texture features.

Figure 4.11 shows the application of the despeckle filter *DsFwaveltc* on a phantom carotid artery image for 10 iterations. Figure 4.4l shows the results of the filter on an artificial carotid image, whereas Figure 4.6k presents the results on a real carotid ultrasound image.

* * * *

CHAPTER 3
Evaluation Methodology

In this chapter, we present the material, the ultrasound imaging scanners used for the image acquisition, as well as the texture analysis, the distance measures, the univariate statistical analysis, and the k-nearest-neighbor (kNN) classifier, which are used to evaluate despeckle filtering. Also, a number of image quality metrics are presented for evaluating the quality between the original and despeckled images. Finally, the procedure of visual evaluation carried out by the experts is introduced.

3.1 MATERIAL AND RECORDING OF ULTRASOUND IMAGES

Two different imaging datasets were used in this study. The first imaging dataset was used for evaluating despeckle filtering, and the second was used for evaluating the image quality of two ultrasound scanners.

The images of the carotid artery bifurcation used for the despeckling (the first image dataset) were acquired using the ATL HDI-3000 ultrasound scanner. The ATL HDI-3000 ultrasound scanner is equipped with a 64-element fine-pitch high-resolution 38-mm broadband array, a multi-element ultrasound scan head with an operating frequency range of 4–7 MHz, an acoustic aperture of 10 × 8 mm, and a transmission focal range of 0.8–11 cm [39]. All images were recorded as they are displayed in the ultrasound monitor after logarithmic compression. The images were digitally recorded on a magneto-optical drive, with a resolution of 768 × 756 pixels with 256 gray levels. The image resolution was 16.66 pixels/mm. B-mode scan settings were adjusted so that the maximum dynamic range was used with a linear postprocessing curve. The position of the probe was adjusted so that the ultrasonic beam was vertical to the artery wall. The time gain compensation curve was adjusted (gently sloping) to produce the uniform intensity of echoes on the screen; however, it was vertical in the lumen of the artery where attenuation in blood was minimal, so that echogenicity of the far wall was the same as that of the near wall. The overall gain was set so that the appearance of the plaque was assessed to be optimal, and slight noise appeared within the lumen. It was then decreased so that at least some areas in the lumen appeared to be free of noise (black).

The first image dataset used for despeckle filtering consisted of a total of 440 ultrasound images of the carotid artery bifurcation, 220 asymptomatic and 220 symptomatic, and was acquired with the ATL HDI-3000 scanner. Asymptomatic images were recorded from patients at risk of atherosclerosis in the absence of clinical symptoms, whereas symptomatic images were recorded from patients at risk of atherosclerosis that have already developed clinical symptoms, such as a stroke episode.

The images used for the image quality evaluation (the second image dataset) were acquired with the ATL HDI-5000 ultrasound scanner, which is equipped with a 256-element fine-pitch high-resolution 50-mm linear array and a multielement ultrasound scan head with an extended operating frequency range of 5–12 MHz, and offers real spatial compound imaging. The scanner increases the image clarity using SonoCT imaging by enhancing the resolution and borders, and interface margins are better displayed. Several tests made by the manufacturer showed that the ATL HDI-5000 scanner was superior overall to conventional two-dimensional imaging systems primarily because of the reduction of speckle, contrast resolution, and tissue differentiation, and the image was visually better [39].

The second image dataset consisted of a total of 80 symptomatic B-mode longitudinal ultrasound images used for the image quality evaluation, from identical vessel segments of the carotid artery bifurcation, which were acquired from each ultrasound scanner (from the ATL HDI-3000 and the ATL HDI-5000 scanner). The images were digitally recorded on a magneto-optical drive with a resolution of 768 × 576 pixels with 256 gray levels. These images were recorded at the Institute of Neurology and Genetics, Nicosia, Cyprus, from 32 female and 48 male symptomatic patients aged between 26 and 95 years old, with a mean age of 54 years old. These subjects were at risk of atherosclerosis, which have already developed clinical symptoms such as a stroke or a transient ischemic attack. In addition, 10 symptomatic ultrasound images of the carotid artery representing different types of atherosclerotic carotid plaque formation with irregular geometry typically found in this blood vessel were acquired from each scanner.

3.2 USE OF PHANTOM AND ARTIFICIAL ULTRASOUND IMAGES

Instead of using real or artificial ultrasound images of the carotid artery, artificial phantoms may also be used, which can be designed for special purposes. An example of an ultrasound phantom of the carotid artery image is shown in Figure 3.1a.

Ultrasound phantoms are used to measure and test system resolution, focal zone, system sensitivity, grayscale dynamic range, penetration, and dead zone [109]. Some phantoms contain liquid that mimics blood flow to test Doppler systems. Most ultrasound phantoms are constructed from acrylic, polyurethane, or aluminum to hold water, gel, urethane rubber, or polystyrene with implants to represent cysts, tumors, plaques, or other abnormalities [109]. Most clinicians rely on Doppler ultrasound measurements, but a subject's maximum blood velocity rates can vary by 10–60% even using the same machine due to calibration problems. The same may be observed with the speckle

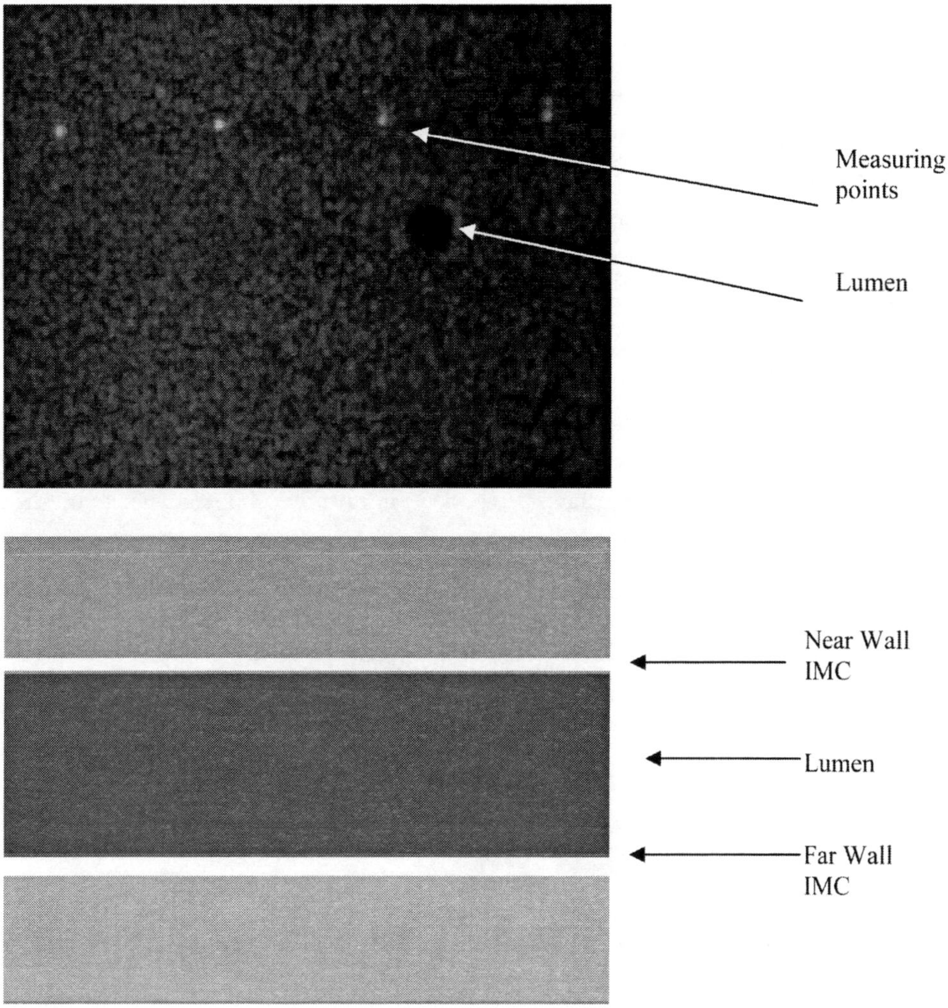

FIGURE 3.1: (a) Phantom ultrasound image. (b) Longitudinal view of an artificial carotid image.

pattern. A different speckle pattern may be obtained from the same patient using the same machine. As a result, it is impossible to obtain an accurate measurement in time. In theory, we can get around this problem if we had independent information on blood flow and speckle pattern through human arteries against which to calibrate the (Doppler) ultrasound data. Nevertheless, to get truly independent and accurate information, in practice, is difficult, and, therefore, the use of the phantoms is required [127].

To further evaluate despeckle filtering, an artificial carotid image was generated (see Figure 3.1b). Despeckle filtering was visually evaluated by two experts (a cardiovascular surgeon and a neurovascular specialist) on the artificial carotid image corrupted by speckle noise. The artificial image (shown in Figure 3.1b) has a resolution of 150 × 150 pixels and was generated with gray-level values of the bottom, strip, middle, and upper segments of 182, 250, 102, and 158, respectively. This image was corrupted by speckle noise (see Figure 4.1a), which was generated using the equation $g_{i,j} = f_{i,j} + n_{i,j} f_{i,j}$, where $g_{i,j}$ and $f_{i,j}$ are the noisy and original images, respectively, and $n_{i,j}$ is uniformly distributed random noise with mean 0 and variance $^2_n = 0.07$.

3.2.1 Types of Plaques

Plaques may be classified into the following types: 1) type I—uniformly echolucent (black), where bright areas occupy less than 15% of the plaque area; 2) type II—predominantly echolucent, where bright echoes occupy 15–50% of the plaque area; 3) type III—predominantly echogenic, where bright echoes occupy 50–85% of the plaque area; 4) type IV—uniformly echogenic, where bright echoes occupy more than 85% of the plaque area; and 5) type V—calcified cap with acoustic shadow so that the rest of the plaque cannot be visualized [53, 60]. In this study, the plaques delineated were of types II, III, and IV because it is easier to make a manual delineation since the fibrous cap, which is the border between blood and plaque, is more easily identified. If the plaque is of type I, borders are not well visible. Plaques of type V produce acoustic shadowing, and the plaque is also not well visible.

3.3 IMAGE NORMALIZATION

The need for image standardization or postprocessing has been suggested in the past, and normalization using only blood echogenicity as a reference point has been applied in ultrasound images of the carotid artery [53]. Brightness adjustments of the ultrasound images have been used in this book, as this has been shown to improve image compatibility by reducing the variability introduced by different gain settings and facilitate ultrasound tissue comparability [53, 93].

The images used for the image quality evaluation were manually normalized by linearly adjusting the image so that the median gray-level value of the blood was 0–5, and the median gray level of the adventitia (the artery wall) was 180–190. The scale of the gray level of the images ranged from 0 to 255 [94].

This normalization using blood and adventitia as reference points was necessary to extract comparable measurements in case of processing images obtained by different operators or different equipment [94]. The image normalization procedure is illustrated in Figure 3.2 and was implemented in MATLAB™ software (version 6.1.0.450, release 12.1, May 2001, by The Mathworks, Inc.) and tested on a Pentium III desktop computer, running at 1.9 GHz with 512-MB RAM.

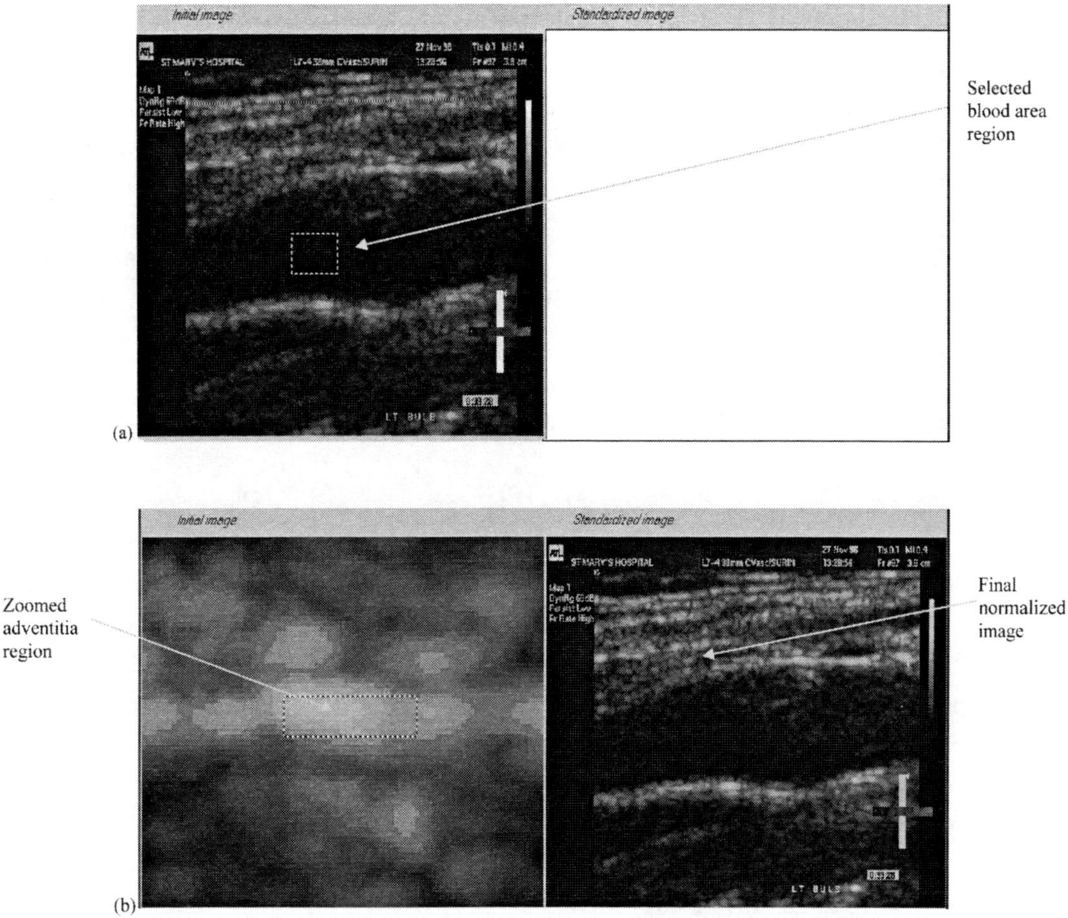

FIGURE 3.2: Normalization of a carotid ultrasound image. Two reference points are selected to normalize the image: (a) blood area and (b) adventitia area.

The same software and computer station were also used for all other methods employed in this book.

3.4 DESPECKLE FILTERING

Ten despeckle filters were selected and investigated, as presented in Chapter 2, and were applied on the 440 logarithmically compressed ultrasound images, artificial carotid images, and phantom carotid images. These filters are (see also Table 1.1) *DsFlsmv*, *DsFlsminsc*, *DsFmedian*, and *DsFwiener* from the linear filtering group; *DsFhomog*, *DsFgf4d*, and *DsFhomo* from the nonlinear filtering group; *DsFad* and *DsFnldif* from the diffusion filtering group; and the *DsFwaveltc* wavelet despeckle filter.

3.5 TEXTURE ANALYSIS

Following the despeckling, texture features may be extracted from the original and despeckled images. Texture analysis is one of the most important features used in image processing and pattern recognition. It can provide information about the arrangement and spatial properties of fundamental image elements. Texture provides useful information for the characterization of atherosclerotic plaque [40]. In this study, a total of 56 different texture features were extracted both from the original and despeckled ultrasound images as follows [40, 41]:

1. *Statistical features* (SF): (a) mean; (b) median; (c) variance (^{2}); (d) skewness (^{3}); (e) kurtosis (^{4}); and (f) speckle index ($^2/m$).
2. *Spatial gray-level dependence matrices* (SGLDM) as proposed by Haralick et al. [41]: (a) angular second moment (ASM); (b) contrast; (c) correlation; (d) sum of squares, variance; (e) inverse difference moment (IDM); (f) sum average; (g) sum variance; (h) sum entropy; (i) entropy; (j) difference variance; (k) difference entropy; (l) and (m) information measures of correlation. Each feature was computed using a distance of one pixel. Also, for each feature, the mean values and the range of values were computed and were used as two different feature sets.
3. *Gray-level difference statistics* (GLDS) [42]: (a) contrast; (b) ASM; (c) entropy; and (d) mean.
4. *Neighborhood gray tone difference matrix* (NGTDM) [43]: (a) coarseness; (b) contrast; (c) business; (d) complexity; and (e) strength.
5. *Statistical feature matrix* (SFM) [44]: (a) coarseness; (b) contrast; (c) periodicity; and (d) roughness.
6. *Laws texture energy measures* (TEM) [44]: For the laws TEM extraction, vectors of length $l = 7$ $L = (1, 6, 15, 20, 15, 6, 1)$, $E = (\ 1, \ 4, \ 5, 0, 5, 4, 1)$, and $S = (\ 1, \ 2, 1, 4, 1, \ 2, \ 1)$ were used, where L performs local averaging, E acts as an edge detector, and S acts as a spot detector. The following TEM features were extracted: (a) LL—texture energy (TE) from LL kernel; (b) EE—TE from EE kernel; (c) SS—TE from SS kernel; (d) LE—average TE from LE and EL kernels; (e) ES—average TE from ES and SE kernels; and (f) LS—average TE from LS and SL kernels.
7. *Fractal dimension texture analysis* (FDTA) [44]: Hurst coefficient $H^{(k)}$ for resolutions k=1, 2, 3, 4.
8. *Fourier power spectrum* (FPS) [44]: (a) radial sum; and (b) angular sum.

Some of the most common texture feature algorithms that have been used for ultrasound texture analysis are simple SF, SGLDM [41], GLDS [42], NGTDM [43], SFM [44], laws TEM [44], FDTA [40, 44], and FPS [42]. These texture features are usually computed on a region of interest, for example, the region prescribed by the plaque contour that is automatically or manually drawn.

The SF includes the , median, ^{2}, ^{3}, and ^{4} values. The SGLDM texture features, as proposed by Haralick et al. [41], are the most frequently used texture features. These are based on the estimation of the second-order joint conditional probability density functions, where two pixel pairs (k, l) and (m, n), with distance d in the direction specified by the angle , have intensities of gray level g and gray level f. Based on the probability density functions, the following texture measures and their variants [128] are computed: ASM, contrast, correlation, IDM, sum average, variance (sum and difference), and entropy (sum and difference). For a chosen distance d that is usually one pixel and for angles $= 0°, 45°, 90°$, and $135°$, four values for each of the above texture measures are computed. The mean and the range of these four values are usually computed for each feature, and they are used as two different feature sets. The GLDS algorithm [42, 80] uses first-order statistics of local property values based on the absolute differences between pairs of gray levels or of average gray levels to extract the following texture measures: contrast, ASM, entropy, and mean. Amadasun and King [43] proposed the NGTDM to extract textural features, which correspond to the visual properties of texture. The following features are extracted: coarseness, contrast, busyness, complexity, and strength. The FDTA feature set is based on the work of Mandelbrot [80], who developed the fractional Brownian motion model to describe the roughness of natural surfaces. The Hurst coefficients ($H^{(k)}$) [44] are computed for different image resolutions, where a smooth texture surface is described by a large value of the parameter $H^{(k)}$, whereas the reverse applies for a rough texture surface. The FPS feature set computes the radial and angular sums of the sample FPS, where coarse texture has high values that are concentrated near the origin, and in fine texture, the values are more spread out.

3.6 DISTANCE MEASURES

To identify the most discriminant features separating the two classes under investigation, i.e., asymptomatic and symptomatic ultrasound images (identifying features that have the highest discriminatory power), before and after despeckle filtering, the distance between asymptomatic and symptomatic images was calculated for the set of all ultrasound images before and after despeckle filtering for each feature as follows [40]:

$$\mathrm{dis}_{zc} = |m_{za} \quad m_{zs}| / \sqrt{\ ^{2}_{za} + \ ^{2}_{zs}}, \tag{3.1}$$

where z is the feature index, c if o indicates the original image set and if f indicates the despeckled image set, m_{za} and m_{zs} are the mean values, and $_{za}$ and $_{zs}$ are the standard deviations of the asymptomatic and symptomatic classes, respectively. The most discriminant features are the ones with the highest distance values [40]. If the distance after despeckle filtering is increased, i.e.,

$$\mathrm{dis}_{zf} > \mathrm{dis}_{zo} \qquad (3.2)$$

then it can be derived that the classes may be better separated.

For each feature, a percentage distance was computed as

$$\mathrm{feat\_dis}_{z} = (\mathrm{dis}_{zf} \quad \mathrm{dis}_{zo})100. \qquad (3.3)$$

For each feature set, a score distance was computed as

$$\mathrm{Score\_Dis} = (1/N) \sum_{z=1}^{N} (\mathrm{dis}_{zf} \quad \mathrm{dis}_{zo})100, \qquad (3.4)$$

where N is the number of features in the feature set. It should be noted that, for all features, a larger feature distance shows improvement.

3.7 UNIVARIATE STATISTICAL ANALYSIS

The Wilcoxon rank-sum test for paired samples is a nonparametric alternative for the paired samples t test when the distribution of the samples is not normal. The Wilcoxon test for paired samples ranks the absolute values of the differences between the paired observations in sample 1 and sample 2, and calculates a statistic on the number of negative and positive differences. If the resulting p value is small ($p < 0.05$), then it can be accepted that the median of the differences between the paired observations is statistically significantly different from 0. The Wilcoxon matched-pairs signed rank-sum test was used to detect if, for each texture feature, a significant (S) or nonsignificant (NS) difference exists between the original and despeckled images at $p < 0.05$. The test was applied on all the 220 asymptomatic and 220 symptomatic original and despeckled images of the carotid artery.

3.8 KNN CLASSIFIER

The statistical pattern recognition kNN classifier using the Euclidean distance with $k = 7$ was used to classify a plaque as asymptomatic or symptomatic [40]. The kNN classifier was chosen because it is simple to implement and is computationally very efficient. This is highly desired due to the many feature sets and filters tested [44]. In the kNN algorithm, to classify a new pattern, its kNNs from the training set are identified. The new pattern is classified to the most frequent class among its neighbors based on a similarity measure that is usually the Euclidean distance. In this work, the kNN carotid plaque classification system was implemented for values of $k = 1, 3, 5, 7$, and 9 using for input the eight texture feature sets and morphology features described above.

The leave-one-out method was used for evaluating the performance of the classifier, where each case is evaluated in relation to the rest of the cases. This procedure is characterized by no bias concerning the possible training and evaluation bootstrap sets. This method calculates the error or the classification score by using n 1 samples in the training set and testing or evaluating the performance of the classifier on the remaining sample. It is known that for large n, this method is computationally expensive. However, it is approximately unbiased at the expense of an increase in the variance of the estimator [104]. The kNN classifier was chosen because it is simple to implement and is computationally very efficient. This is highly desired due to the many feature sets and filters tested [44].

3.9 IMAGE QUALITY EVALUATION METRICS

For medical images, quality can be objectively defined in terms of performance in clinically relevant tasks such as lesion detection and classification, where typical tasks are the detection of an abnormality, the estimation of some parameters of interest, or the combination of the above [52]. Most studies today have assessed the equipment performance by testing diagnostic performance of multiple experts, which also suffer from intraobserver and interobserver variability. Although this is the most important method of assessing the results of image degradation, few studies have attempted to perform physical measurements of degradation [47]. Image quality is important when evaluating or segmenting atherosclerotic carotid plaques [60] or the intima media complex (IMC) in the carotid artery [54], where speckle obscures subtle details [7] in the image. In a recent study [59], we have shown that speckle reduction improves the visual perception of the expert in the assessment of ultrasound imaging of the carotid artery.

To be able to design accurate and reliable image quality metrics, it is necessary to understand what quality means to the expert. An expert's satisfaction when watching an image depends on many factors. One of the most important is, of course, image content. Research made in the area of image quality showed that this depends on many parameters, such as viewing distance, display size, resolution, brightness, contrast, sharpness, colorfulness, naturalness, and other factors [90].

It is also important to note that there is often a difference between fidelity (the accurate reproduction of the original on the display) and perceived quality. Sharp images with high contrast are usually more appealing to the average expert. Likewise, subjects prefer slightly more colorful and saturated images despite realizing that they look somewhat unnatural [92]. For studying visual quality, some of the definitions above should be related to the human visual system. Unfortunately, subjective quality may not be described by an exact figure; due to its inherent subjectivity, it can only be statistically described. Even in psychological threshold experiments, where the task of the expert is to give a yes or no answer, there exists a significant variation between expert's contrast sensitivity functions and other critical low-level visual parameters. When speckle noise is apparent in the image, the expert's differing experiences with noise are bound to lead to different weightings of the artifact [90]. Researchers showed that experts and nonexperts examine different critical image characteristics to form their final opinion with respect to image quality [91]. Thus, image quality evaluation metrics can be used for the evaluation of despeckle filtering.

Differences between the original ($g_{i,j}$) and despeckled ($f_{i,j}$) images were evaluated using image quality evaluation metrics. The following measures, which are easy to compute and have clear physical meaning, were computed.

1. The MSE

$$\mathrm{MSE} = \frac{1}{MN} \sum_{i=1}^{M}\sum_{j=1}^{N} (g_{i,j} \quad f_{i,j})^2, \tag{3.5}$$

   which measures the quality change between the original and processed images in an $M \times N$ window [45]. The MSE has been widely used to quantify image quality, and, when it is used alone, it does not correlate strongly enough with perceptual quality. It should be used, therefore, together with other quality metrics and visual perception [45, 48].

2. The root MSE (RMSE), which is the square root of the squared error averaged over an $M \times N$ window [46]:

$$\mathrm{RMSE} = \sqrt{\frac{1}{MN} \sum_{i=1}^{M}\sum_{j=1}^{N} (g_{i,j} \quad f_{i,j})^2}. \tag{3.6}$$

   The popularity of the RMSE arises mostly from the fact that it is, in general, the best approximation of the standard error.

3. The error summation in the form of the Minkowski metric, which is the norm of the dissimilarity between the original and despeckled images [47]:

$$\mathrm{Err} = \left(\frac{1}{MN} \sum_{i=1}^{M}\sum_{j=1}^{N} |g_{i,j} \quad f_{i,j}| \right)^{1/} \tag{3.7}$$

   computed for = 3 (Err_3) and = 4 (Err_4). For = 2, the RMSE is computed as in Eq. (3.6), whereas for = 1, the RMSE is computed as the absolute difference, and for = , the RMSE is computed as the maximum difference measure.

4. The geometric average error (GAE) is a measure that shows that if the despeckled image is very bad, it is used to replace or complete the RMSE and is computed as follows [48]:

$$\mathrm{GAE} = \left(\prod_{i=1}^{M} \prod_{j=1}^{N} \sqrt{g_{i,j} \quad f_{i,j}} \right)^{1/MN}. \tag{3.8}$$

The value of GAE is approaching zero if there is a very good transformation (small differences) between the original and despeckled images; otherwise, the value of GAE is high. It is positive only if every pixel value is different between the original and despeckled images. This measure is also used for teleultrasound when transmitting ultrasound images. The GAE may be used to replace the RMSE, which is dominated by its large individual terms and is calculated for an image with dimensions $N \times M$. This amounts to a severe error in the RMSE when large individual terms are present. For this reason, the RMSE is often replaced by the GAE.

5. Although signal sensitivity and image noise properties are important by themselves, it is really their ratio that carries the most significance. The signal-to-noise ratio (SNR) is given by [49]

$$\mathrm{SNR} = 10\log_{10}\frac{\sum_{i=1}^{M}\sum_{j=1}^{N}(g_{i,j}^2 \quad f_{i,j}^2)}{\sum_{i=1}^{M}\sum_{j=1}^{N}(g_{i,j} \quad f_{i,j})^2} \tag{3.9}$$

It is calculated over an image area with dimensions $N \times M$. The SNR, the RMSE, and the Err proved to be very sensitive tests for image degradation, but they are completely nonspecific. Any small change in image noise, despeckling, and transmitting preferences would cause an increase in the above measures.

6. The peak SNR (PSNR) is computed using [49]

$$\mathrm{PSNR} = 10\log_{10}\frac{\mathrm{MSE}}{g_{\max}^2}, \tag{3.10}$$

where $g_{\max}^2$ is the maximum intensity in the unfiltered image. The PSNR is higher for a better transformed image and lower for a poorly transformed image. It measures image fidelity, which is how closely the despeckled image resembles the original image.

7. The mathematically defined universal quality index [50] models any distortion as a combination of three different factors, which are loss of correlation, luminance distortion, and contrast distortion, and is derived as

$$Q = \frac{\sigma_{gf}}{\sigma_f \sigma_g} \cdot \frac{2\bar{f}\bar{g}}{(\bar{f})^2 + (\bar{g})^2} \cdot \frac{2\sigma_f \sigma_g}{\sigma_f^2 + \sigma_g^2}, \quad 1 < Q < 1, \tag{3.11}$$

where $\bar{g}$ and $\bar{f}$ represent the mean of the original and despeckled values with their standard deviations σ_g and σ_f of the original and despeckled values of the analysis window, and σ_{gf} represents the covariance between the original and despeckled windows. Q is computed for a sliding window of size 8 × 8 without overlapping. Its highest value is 1 if $g_{i,j} = f_{i,j}$, whereas its lowest value is 1 if $f_{i,j} = 2\bar{g} \quad g_{i,j}$.

8. The structural similarity index (SSIN) between two images [47], which is a generalization of Eq. (3.11), is given by

$$\mathrm{SSIN} = \frac{(2\bar{g}\bar{f} + c_1)(2 \quad_{gf} + c_2)}{(\bar{g}^2 + \bar{f}^2 + c_1)(\ ^2_g + \ ^2_f + c_2)}, \quad 1 < \mathrm{SSIN} < 1, \tag{3.12}$$

where $c_1 = 0.01dr$ and $c_2 = \mathbf{0.03}dr$, with $dr = 255$ representing the dynamic range of the ultrasound images. The range of values for the SSIN lies between −1 for a bad similarity and 1 for a good similarity between the original and despeckled images, respectively. It is similarly computed to the Q measure for a sliding window of size 8 × 8 without overlapping.

9. The speckle index C for log-compressed ultrasound images is defined as

$$C = \frac{1}{MN} \sum_{i=1}^{M} \sum_{j=1}^{N} \frac{\ ^2_{i,j}}{\ _{i,j}} \tag{3.13}$$

and is an average measure of the amount of speckle presented in the image area with size $M \quad N$ as a whole (over the whole image). It is used in most adaptive filters to adjust the weighting function $k_{i,j}$ in Eq. (2.2) described in Chapter 2 because it reflects the changes in contrast of the image in the presence of speckle noise. It does not depend on the intensity of the local mean, but on the variance $\ ^2$ and the mean $\quad$ of the whole image. The larger the value of C is, the more likely that the observed neighborhood belongs to an edge; thus, C may be also used as an edge detector.

10. Lesion detectability can be quantified using the contrast-to-speckle ratio (CSR) [31, 56]. It is calculated by defining two regions of interest (i.e., the original image and the despeckled) and using the mean pixel value and the variance to quantify the contrast $(\ _1 \quad _2)/\ _1$ and the speckle index noise $\sqrt{(\ ^2_1 + \ ^2_2)/\ _1}$. The ratio of these two quantities is termed as CSR and is defined as

$$\mathrm{CSR} = ((\ _1 \quad _2)\ _1)/\sqrt{(\ ^2_1 + \ ^2_2)}, \tag{3.14}$$

where $\ _1$, $\ _2$, $\ _1$, and $\ _2$ are the mean and standard deviations of the original and despeckled images, respectively. The CSR provides a quantitative measure of the detectability of low contrast lesions, when one region is completely inside the lesion, and the second is the background media.

It is noted that a new image quality metric based on natural scene statistics and mutual information between the original and filtered images has recently been proposed by Sheikh et al. [51].

The quality measures proposed above do not necessarily correspond to all aspects of the expert's visual perception of the errors, nor do they correctly reflect structural coding artifacts [59]. However, if they are all combined together, and with the subjective tests, they may offer a more accurate evaluation result. Subjective tests are tedious, time consuming, and expensive, and the results depend on the expert's background, motivation, and other factors [47, 48, 74, 76]. However, all these measures cover the visual quality just partly. The visual quality of an image is difficult to define with mathematical precision since it is dependent on the properties of our visual system. We know, for example, that our visual system is more tolerant to a certain amount of noise than to a reduced sharpness. On the other hand, it is very sensitive to certain specific artifacts, like blips and bumps [52].

We have, furthermore, investigated in this book the usefulness of image quality evaluation based on the above proposed image quality metrics and visual perception evaluation by experts (see Chapter 4) in ultrasound imaging of the carotid artery after normalization and speckle reduction filtering. For this task, we have evaluated the quality of ultrasound imaging of the carotid artery on two different ultrasound scanners—the HDI ATL-3000 and the HDI ATL-5000—before and after speckle reduction, after image normalization, and after image normalization and speckle reduction filtering. Statistical and texture analysis was carried out on the original and processed images, and the findings were compared with the visual perception carried out by two experts.

3.10 VISUAL EVALUATION BY EXPERTS

Visual evaluation can be broadly categorized as the ability of an expert to extract useful anatomical information from an ultrasound image. The visual evaluation varies, of course, from expert to expert and is subject to the observer's variability [52]. The visual evaluation was carried out according to the International Telecommunication Union Radiocommunication Sector recommendations with the double-stimulus continuous quality-scale procedure [48].

For the despeckle filtering evaluation, a total of 100 ultrasound images of the carotid artery bifurcation (50 asymptomatic and 50 symptomatic) were visually evaluated by two vascular experts—a cardiovascular surgeon and a neurovascular specialist—before and after despeckle filtering. For each case, the original and despeckled images (despeckled with filters *DsFlsmv, DsFlsminsc, DsFmedian*, *DsFwiener*, *DsFhomog*, *DsFgf4d*, *DsFhomo*, *DsFad*, *DsFnldif*, and *DsFwaveltc*) were presented without labeling at random to the two experts. The experts were asked to assign a score in the 1–5 scale corresponding to low and high subjective visual perception criteria. Five was given to an image with the best visual perception. Therefore, the maximum score for a filter is 500 if the expert assigned a score of 5 for all the 100 images. For each filter, the score was divided by five to be expressed in the percentage format. The experts were allowed to give equal scores to more

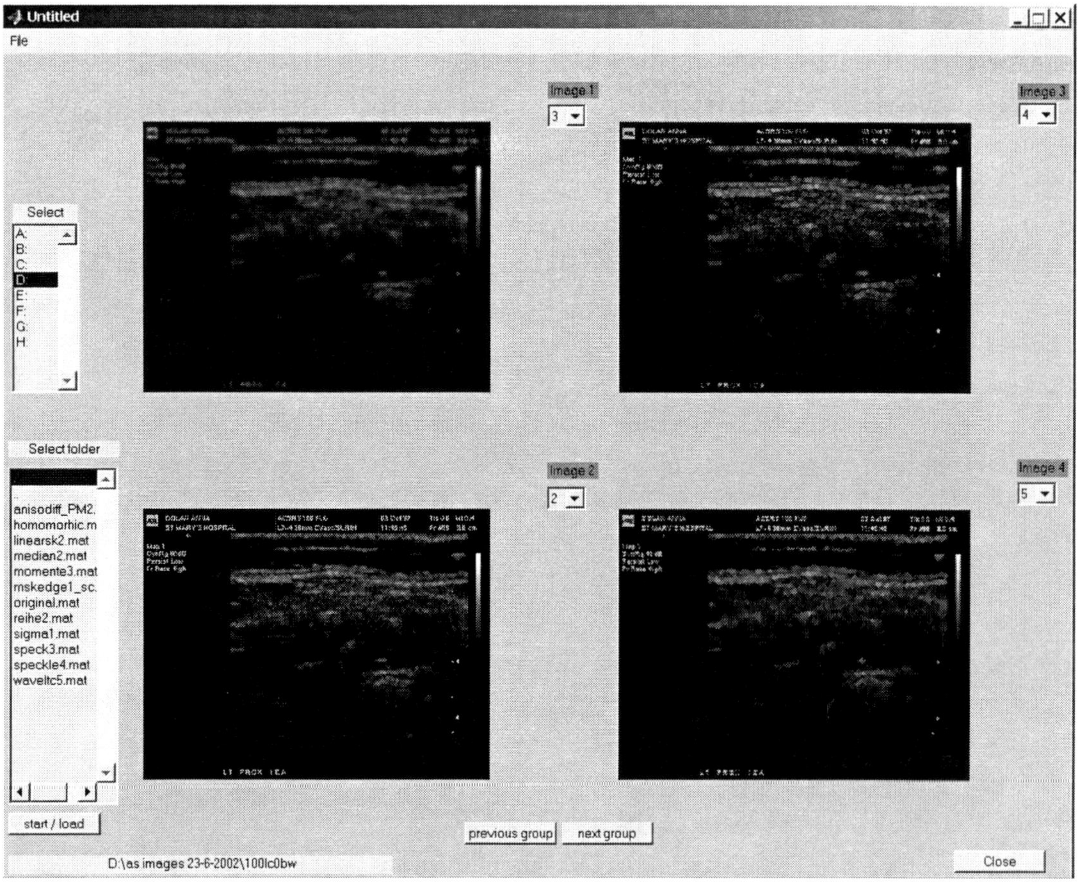

FIGURE 3.3: The graphical user interface for the visual image evaluation carried out by the experts. The screen illustrates four different despeckled images and their corresponding scores.

than one image in each case. For each class and for each filter, the average score was computed. A graphical user interface was developed in MATLAB™, as shown in Figure 3.3, and was used by the two experts for the visual perception evaluation. For each case, the original and despeckled images (despeckled with filters *DsFlsmv*, *DsFlsminsc*, *DsFmedian*, *DsFwiener*, *DsFls*, *DsFhomog*, *DsFgf4d*, *DsFhomo*, *DsFad*, *DsFnldif*, and *DsFwaveltc*) were presented without labeling at random to the two experts.

All the visual evaluation experiments were carried out at the same workstation under indirect fluorescent lighting typical of an office environment. The two vascular experts were allowed to comfortably position themselves with respect to the viewing monitor, where a typical distance of about 50 cm was kept. Experts in real-life applications employ a variety of conscious and unconscious strategies for image evaluation, and it was our intent to create an environment as close as possible to the real one. The two vascular experts evaluated the area around the distal common carotid 2–3 cm before the bifurcation and the bifurcation. It is known that measurements taken from the far wall of the carotid artery are more accurate than those taken from the near wall [53]. Furthermore, the experts examined the image in the lumen area to identify whether a plaque existed.

To further assess the intraobserver variability, the two experts evaluated the same set of images approximately one year after the initial evaluation.

For the image quality evaluation, two vascular experts evaluated the images. The vascular experts—an angiologist and a neurovascular specialist—were allowed to comfortably position themselves with respect to the viewing monitor, where a typical distance of about 50 cm was kept. Experts in real-life applications employ a variety of conscious and unconscious strategies for image evaluation, and it was our intent to create an application environment as close as possible to the real one. The two vascular experts evaluated 80 ultrasound images recorded from each ultrasound scanner before and after speckle reduction, after image normalization, and after normalization and speckle reduction filtering.

The two vascular experts evaluated the area around the distal common carotid between 2 and 3 cm before the bifurcation and the bifurcation. It is known that measurements taken from the far wall of the carotid artery are more accurate than those taken from the near wall [53, 54, 59, 93, 94]. Furthermore, the experts examined the image in the lumen area to identify whether a plaque existed. The primary interest of the experts was the area around the borders between blood and tissue of the carotid artery, and how much better they can differentiate blood from carotid wall, intima media, or plaque surface.

For each image, an individual expert is asked to assign a score in the 1–5 scale corresponding to low and high subjective visual perception criteria. Five was given to an image with the best visual perception. Therefore, the maximum score for a procedure is 400 if the expert assigned a score of 5 for all the 80 images. For each procedure, the score was divided by four to be expressed in the percentage format. The experts were allowed to give equal scores to more than one image in each case. For each preprocessing procedure, the average score was computed.

* * * *

CHAPTER 4

Applications of Despeckle Filtering in Ultrasound Imaging

In this chapter, we first present the evaluation of despeckle filtering on phantom and artificial ultrasound images. Furthermore, the evaluation of the despeckle filters described in Chapter 2 is applied on 220 asymptomatic and 220 symptomatic ultrasound images of the carotid artery bifurcation. A total of 56 texture features were computed, and the most discriminant ones are presented. The performance of these filters is investigated for discriminating between asymptomatic and symptomatic images using the statistical kNN classifier. Moreover, nine different image quality evaluation metrics were computed, as well as visual evaluation scores carried out by two experts.

In addition, we present the image quality evaluation results performed on 80 ultrasound images of the CCA, which were acquired from two different ultrasound imaging scanners (ATL HDI-3000 and ATL HDI-5000). Texture features and image quality evaluation metrics were also extracted from the original and the despeckled images from both scanners, and two experts evaluated the images visually.

Furthermore, we present results from the segmentation of the intima–media complex, the segmentation of the atherosclerotic plaque, after the application of despeckle filtering and normalization. Finally, some results on video despeckling are presented.

4.1 EVALUATION OF DESPECKLE FILTERING ON PHANTOM AND ARTIFICIAL IMAGES

4.1.1 Phantom Image

Despeckle filtering was evaluated on a phantom carotid artery image (see Figure 3.1a). Figure 4.1 shows the despeckled phantom ultrasound images of the carotid artery after the application of the despeckle filters *DsFlsmv*, *DsFlsminsc*, *DsFwiener*, *DsFmedian*, *DsFls*, *DsFhomog*, *DsFgf4d*, *DsFhomo*, *DsFad*, *DsFsrad*, *DsFnldif*, and *DsFwaveltc* for different pixel moving window size and

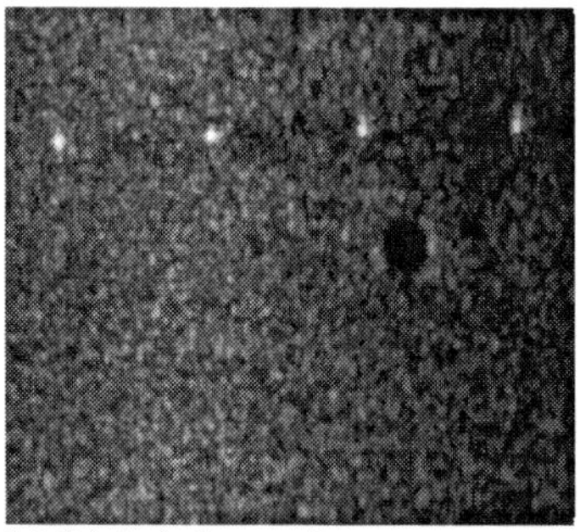

(a) Original phantom image

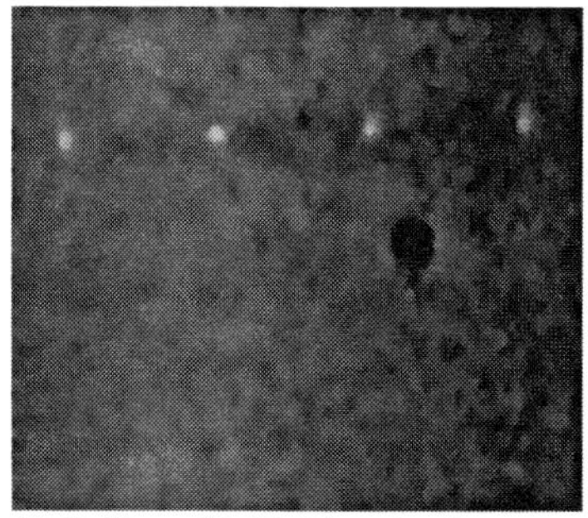

(b) *DsFlsmv* ([7x7], 5 iterations)

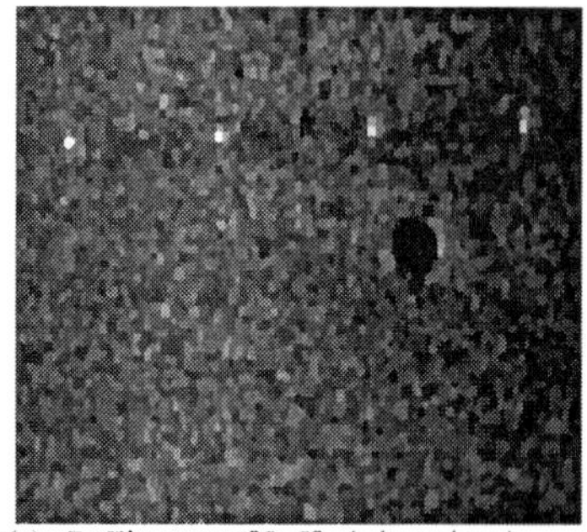

(c) *DsFlsminsc* ([5x5], 4 iterations)

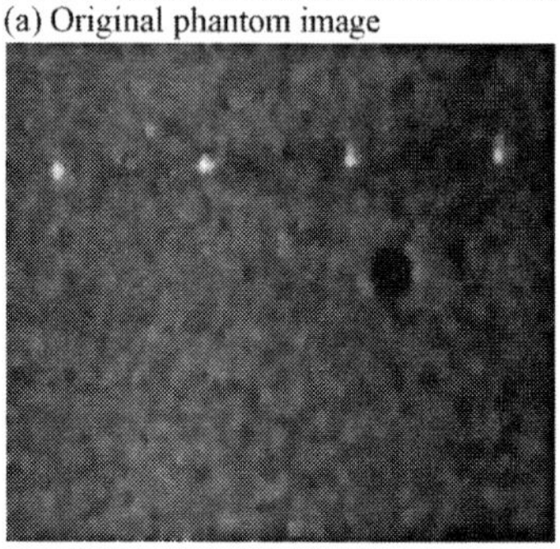

(d) *DsFwiener* ([5x5], 4 iterations)

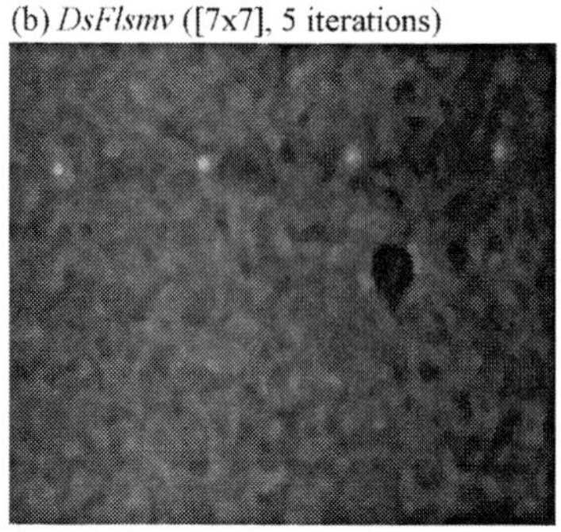

(e) *DsFmedian* ([5x5], 4 iterations)

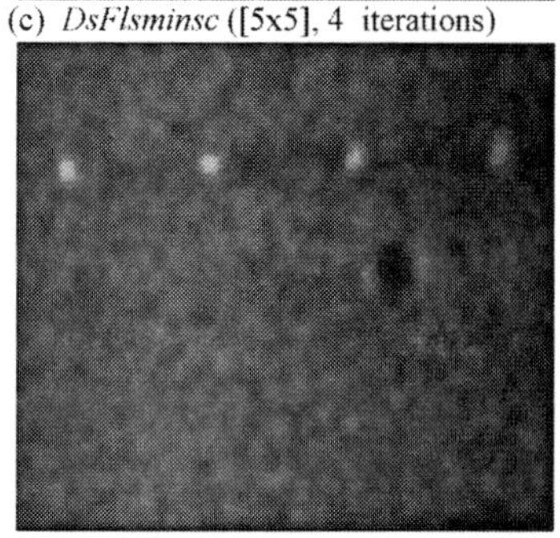

(f) *DsFls* ([5x5] 3 iterations)

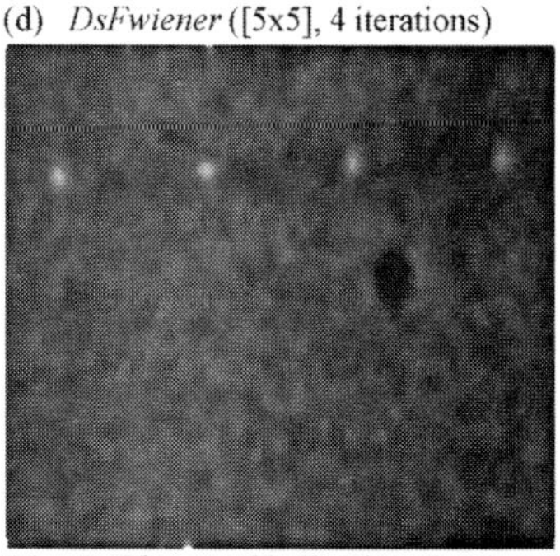

(g) *DsFhomog* ([7x7], 5 iterations)

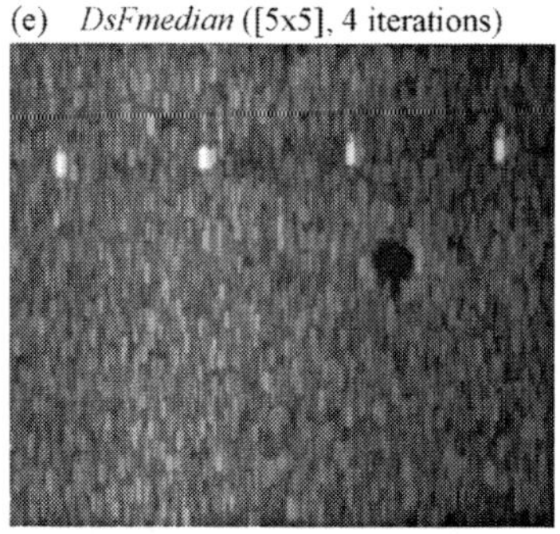

(h) *DsFgf4d* ([7x7], 6 iterations)

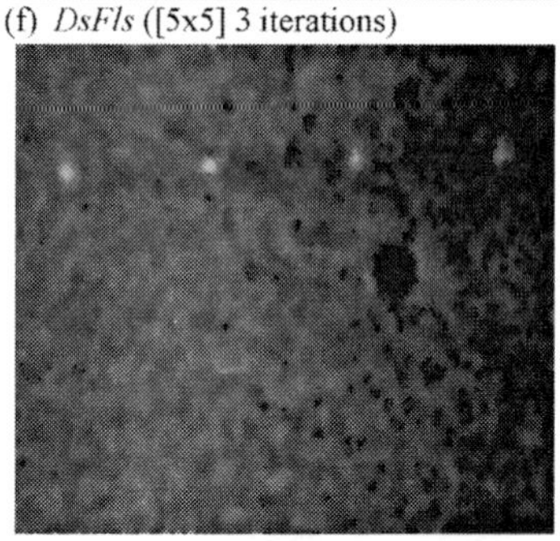

(i) *DsFhomo* ([5x5], 4 iterations)

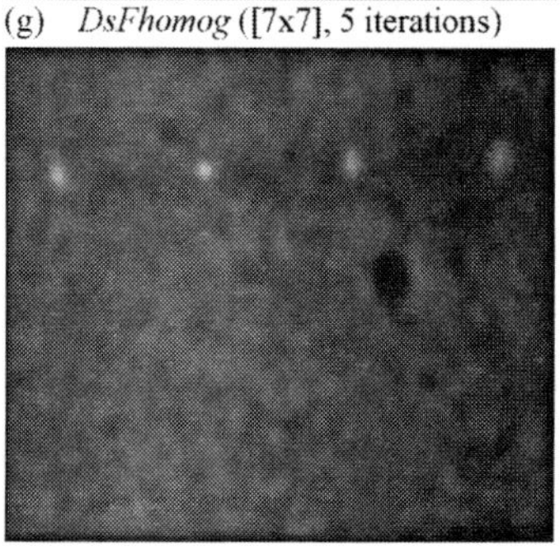

(j) *DsFad* (30 iterations)

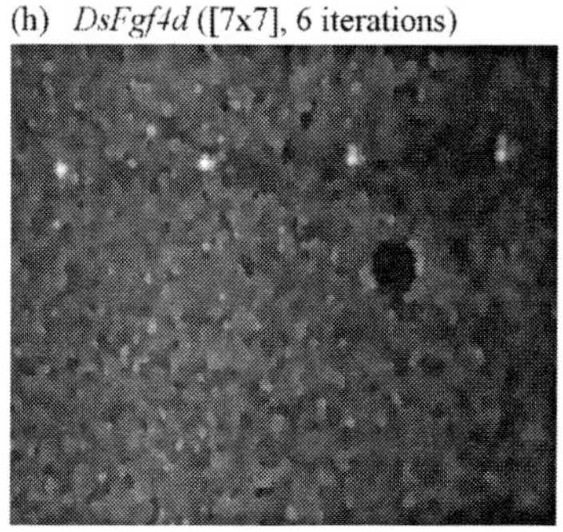

(k) *DsFnldif* (20 iterations)

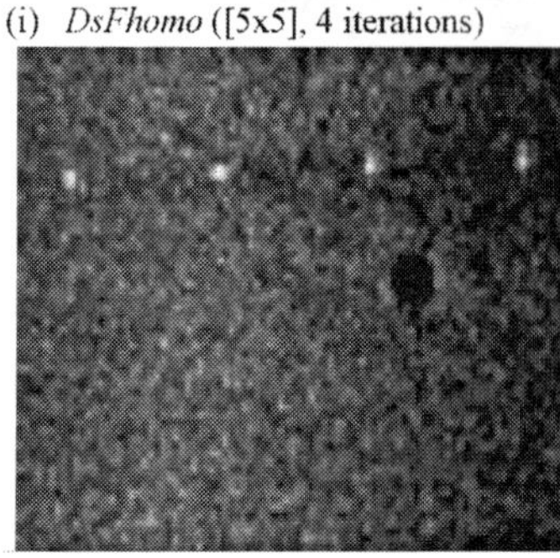

(l) *DsFwaveltc* (10 iterations)

number of iterations (shown in parentheses). Best results were given for the filters *DsFmedian*, *DsFwiener*, *DsFlsmv*, *DsFlsminsc*, and *DsFgf4d*. The filters *DsFad*, *DsFsrad*, *DsFnldif*, *DsFls*, *DsFwaveltc*, *DsFhomog*, and *DsFhomo* did not preserve the edges. Moreover, it is shown from Figure 4.1h and k that the filters *DsFhomo* and *DsFnldif* were noisy.

Table 4.1a tabulates the statistical features, , median, ^{2}, ^{3}, ^{4}, the NGTDM contrast, the speckle index, *C* (3.13), and the contrast-speckle-radio, CSR (3.14), for the phantom image and the 11 filters illustrated in Figure 4.1. As shown in Table 4.1a, all filters, reduced *C* with the exception of the *DsFlsminsc* filter. The CSR is better for the *DsFsrad*, *DsFgf4d*, and *DsFhomo*. Filters that reduced the variance, ^{2}, while preserving the mean, , and the median compared to the original image, were: *DsFhomo*, *DsFls*, *DsFwiener*, *DsFwaveltc*, *DsFad*, *DsFhomog*, *DsFmedian*, and *DsFlsmv*. The contrast, of the image is increased by the filter *DsFlsminsc* (enormously), and preserved by *DsFgf4d*, *DsFsrad*, *DsFwaveltc*, and *DsFnldif*. It is decreased by the filters *DsFlsmv*, *DsFmedian*, *DsFwiener*, *DsFls*, *DsFhomog*, and *DsFad*. It is noted that filters *DsFgf4d*, *DsFlsmv*, and *DsFlsminsc* reduced *C*, *DsFgf4d* increased CSR, *DsFlsmv* reduced the contrast, whereas *DsFlsminsc* increased the contrast. The despeckled images of Figure 4.1 were also assessed by the two experts. Filters that showed an improved smoothing after filtering, as assessed visually by the two experts, using visual perception criteria, are presented in the following order: *DsFwaveltc*, *DsFlsmv*, *DsFnldif*, *DsFsrad*, *DsFad*, *DsFgf4d*, and *DsFmedian*. Filters that showed a blurring effect especially on the edges were: *DsFls*, *DsFlsminsc*, *DsFhomog*, *DsFhomo*, and *DsFwiener*.

Despeckle filtering was furthermore evaluated on the phantom image varying the number of iterations. The despeckle filter *DsFlsmv* was applied on the phantom image for different number of iterations (1 to 15) where the size of the sliding moving window was [7 × 7] as illustrated in Figure 4.2.

Table 4.1b tabulates the same statistical features presented also in Table 4.1a for the phantom image after the application of the *DsFlsmv* despeckle filter. It is shown in Table 4.1b that, for

---

FIGURE 4.1: Original phantom image given in (a), and the application of 11 despeckle filters for different number of iterations and different pixel moving window sizes, shown in brackets, given in (b)–(l). (a) Original phantom image. (b) *DsFlsmv* ([7 × 7], five iterations). (c) *DsFlsminsc* ([5 × 5], four iterations). (d) *DsFwiener* ([5 × 5], four iterations). (e) *DsFmedian* ([5 × 5], four iterations). (f) *DsFls* ([5 × 5] three iterations). (g) *DsFhomog* ([7 × 7], five iterations). (h) *DsFgf4d* ([7 × 7], six iterations). (i) *DsFhomo* ([5 × 5], four iterations). (j) *DsFad* (30 iterations). (k) *DsFnldif* (20 iterations). (l) *DsFwaveltc* (10 iterations).

TABLE 4.1A: Selected statistical features for Figure 4.1 before and after despeckle filtering of all despeckle filters for different number of iterations and different pixel moving window sizes

FEATURE		LINEAR FILTERING				
	ORIGINAL	*DsFlsmv*	*DsFlsminsc*	*DsFwiener*	*DsFmedian*	*DsFls*
NO. OF ITERATIONS		5	4	4	4	3
	36	35	36	36	35	36
Median	37	40	37	39	39	39
^{2}	21	17	22	18	18	16
^{3}	0.2	0.3	0.2	0.06	0.3	0.1
^{4}	2.8	2	3	3	2	2.3
Contrast	76	4	**114**	6	**7**	5
$C = (\ ^2/\ )100$	58	**48**	61	**50**	**51**	**44**
CSR*100		0.5	0	0	0.4	0

Bold values show improvement after despeckle filtering.

increasing number of iterations, the mean, median, skewness, ^{3}, and kurtosis, ^{4}, are preserved, whereas the standard deviation, ^{2}, is reduced. Furthermore, it is shown that, for increasing number of iterations, the filter *DsFlsmv* reduced *C*, whereas an increase in *C* for iterations 7, 8, and 9 was observed. The CSR is better after the seventh iteration of the *DsFlsmv* filter. The despeckled phantom images of Figure 4.2 were also visually assessed by the two experts, where the best visual results were given for iterations 4, 5, and 6.

Despeckle filtering was also evaluated on the phantom image based on the size of the moving window. The despeckle filter *DsFlsmv* was applied on the phantom image for different moving window sizes (from [3 × 3] to [23 × 23]) where the number of iterations was kept constant at 5.

Table 4.1c tabulates the statistical features also presented in Table 4.1a for the phantom image after the application of the *DsFlsmv* filter. It is shown that for increasing sliding moving window size, the mean and median are preserved, the skewness, ^{3}, is increased, and the kurtosis, ^{4}, and the contrast are decreased. The speckle index, *C*, is reduced, while the CSR remains constant.

NONLINEAR FILTERING			DIFFUSION			WAVELET
DsFhomog	*DsFgf4d*	*DsFhomo*	*DsFad*	*DsFsrad*	*DsFnldif*	*DsFwaveltc*
5	6	4	30	50	20	10
37	47	34	35	50	36	36
39	50	38	39	52	39	38
17	22	19	17	27	19	20
0.2	0.02	0.3	0.3	0.2	0.1	0.1
2.2	3	2	2	3	3.1	2.8
3	**50**	13	3	**57**	19	**29**
46	**47**	56	49	54	**53**	**55**
0.4	**4.7**	0.95	0.4	**5.8**	0	0

Furthermore, it is shown that by increasing the number of iterations, the filter *DsFlsmv* reduced the speckle index, *C*, whereas the CSR remains constant. The despeckled phantom images of Figure 4.3 were also visually assessed by the two experts, where the best visual results were given for the sliding window sizes of [3 × 3] and [5 × 5].

4.1.2 Artificial Carotid Image

Despeckle filtering was also evaluated on an artificial carotid artery image corrupted by speckle noise (see Figure 4.4a) as described in Section 3.1. Figure 4.4 shows the original noisy image of the artificial carotid artery, degraded by speckle noise, together with the despeckled images. Figure 4.5 shows line profiles (intensity), for the line marked in Figure 4.4a for all despeckle filters. The profile results show that most of the filters (*DsFmedian*, *DsFwiener*, *DsFlsmv*, *DsFwaveltc*, *DsFlsminsc*, and *DsFgf4d*) preserved the edge boundaries preserving the locality and minimally affecting the reference values in each region. Best results were obtained for the filters *DsFmedian*, *DsFwiener*, *DsFlsmv*, *DsFlsminsc*, and *DsFgf4d*. The filters *DsFad*, *DsFnldif*, *DsFls*, *DsFwaveltc*, *DsFhomog*, and

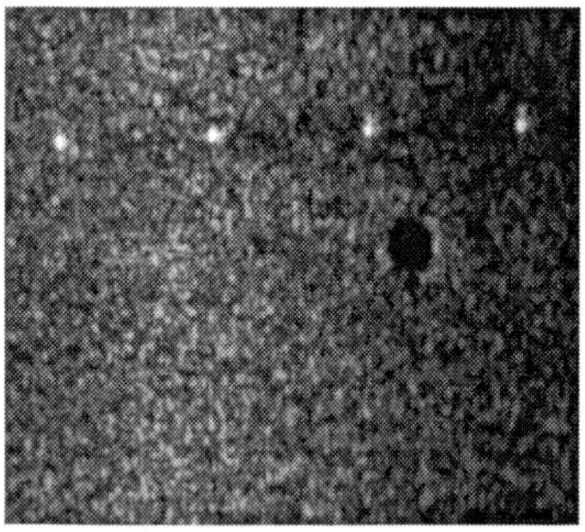
(a) Original

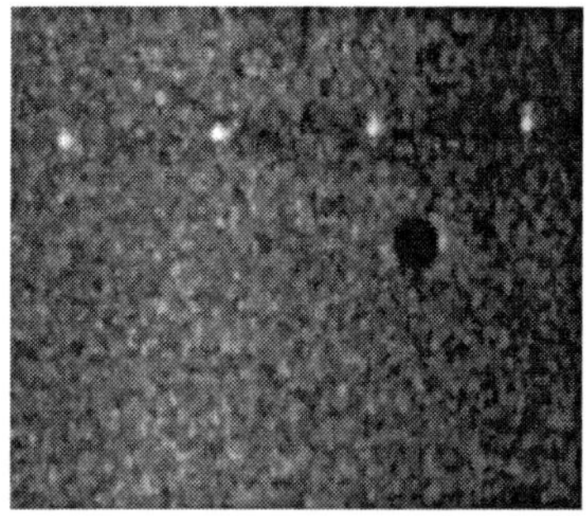
(b) 1 iteration

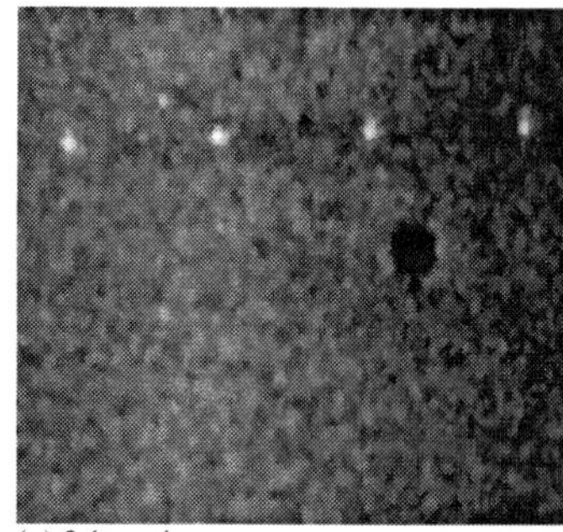
(c) 2 iterations

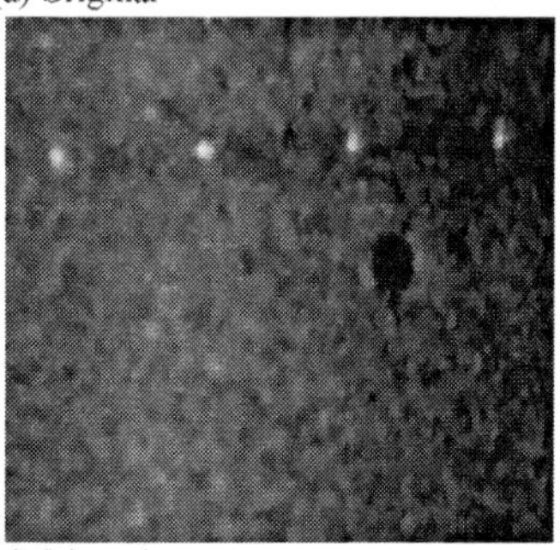
(d) 3 iterations

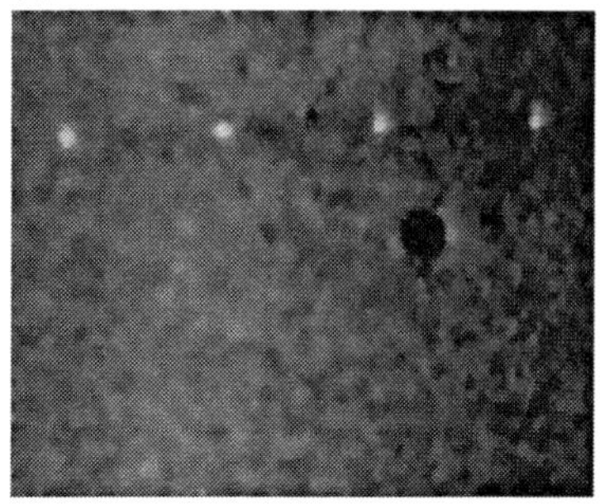
(e) 4 iterations

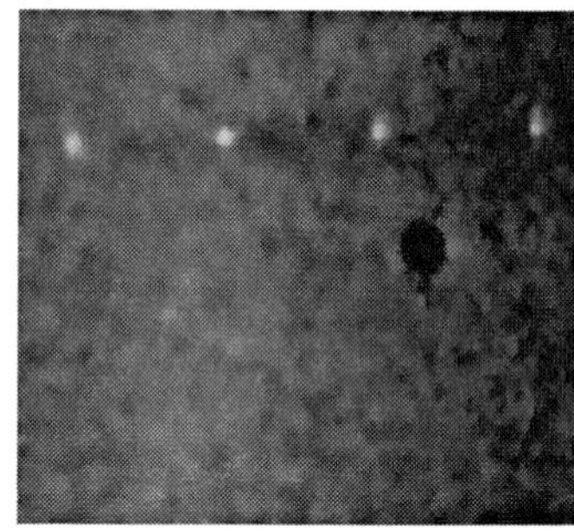
(f) 5 iterations

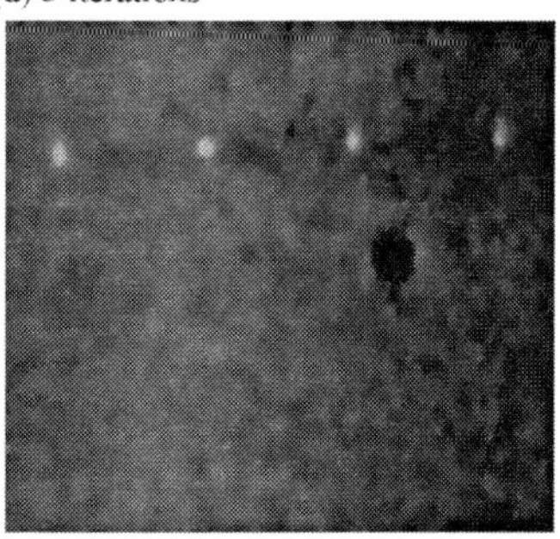
(g) 6 iterations

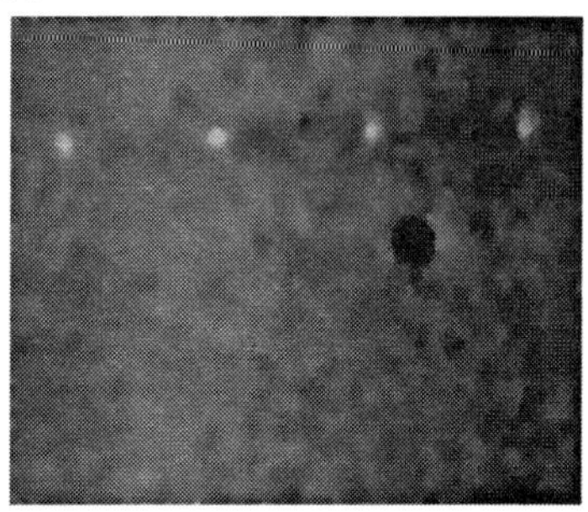
(h) 7 iterations

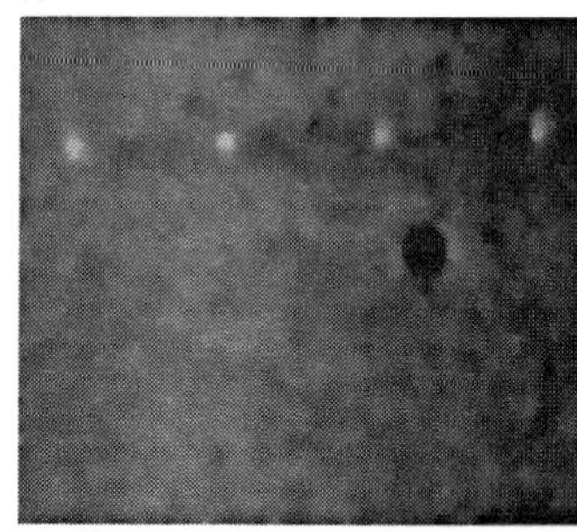
(i) 8 iterations

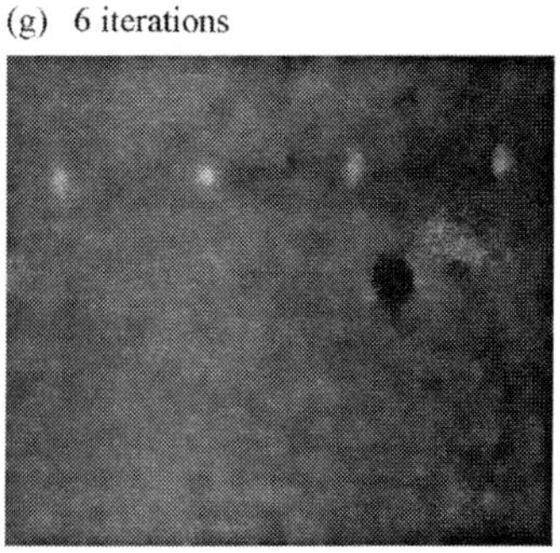
(j) 9 iterations

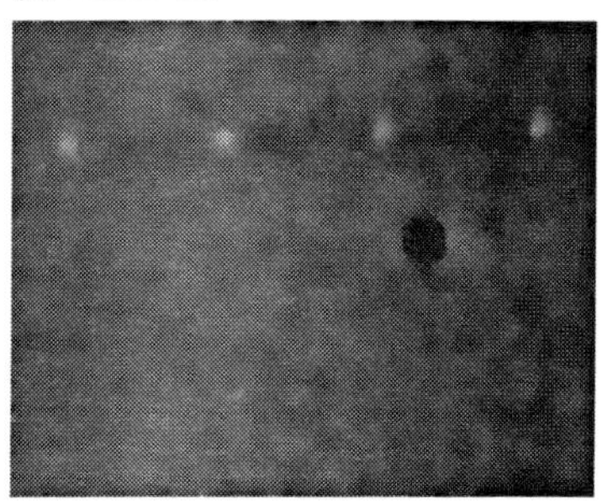
(k) 10 iterations

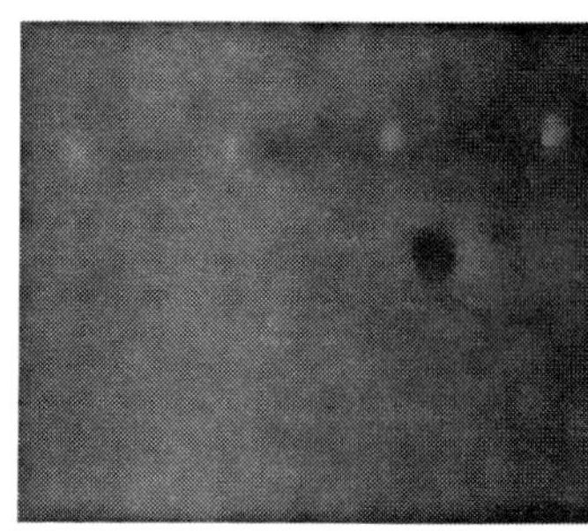
(l) 15 iterations

TABLE 4.1B: Selected statistical features for Figure 4.1 before and after despeckle filtering for increasing number of iterations and constant window size ([7 × 7]) for the despeckle filter *DsFlsmv*

FEATURE	NUMBER OF ITERATIONS												
	ORIGINAL	1	2	3	4	5	6	7	8	9	10	15	20
	36	36	36	36	36	36	36	35	35	35	35	35	35
Median	37	38	39	39	40	40	40	40	40	40	40	40	39
2	21	19	18	18	17	17	17	17	17	17	16	16	16
3	0.2	0.01	0.2	0.2	0.3	0.3	0.3	0.3	0.3	0.4	0.4	0.4	0.4
4	2.8	2.6	2.4	2.3	2.1	2.1	2	2	2	1.9	1.9	1.8	1.8
Contrast	76	35	17	9	6	3.8	2.9	2	1.7	1.4	1.3	0.8	0.6
$C=(\ ^{2}/\ )100$	58	53	50	50	47	47	47	49	49	49	46	46	46
CSR*100		0	0	0	0	0	0	0.45	0.45	0.45	0.46	0.46	0.46

Bold values show improvement after despeckle filtering.

DsFhomo do not preserve the edges, moving the line profiles to darker grayscale values. Moreover, it is shown from Figure 4.5i, that the filter *DsFhomo* is very noisy.

The despeckled images of Figure 4.4 were also assessed by the two experts. Filters that showed an improved smoothing after filtering, as assessed visually by the two experts, using visual perception criteria, are presented in the following order: *DsFwaveltc*, *DsFlsmv*, *DsFnldif*, *DsFad*, *DsFgf4d*, and *DsFmedian*. Filters that showed a blurring effect especially on the edges were: *DsFls*, *DsFlsminsc*, *DsFhomog*, *DsFhomo*, and *DsFwiener*.

FIGURE 4.2: Original phantom image given in (a), and the despeckled phantom images after the application of the *DsFlsmv* filter for increasing number of iterations and a constant pixel moving window size of [7 × 7] given in (b)–(l). The number of iterations ranges from 1 to 15. (a) Original. (b) One iteration. (c) Two iterations. (d) Three iterations. (e) Four iterations. (f) Five iterations. (g) Six iterations. (h) Seven iterations. (i) Eight iterations. (j) Nine iterations. (k) 10 iterations. (l) 15 iterations.

TABLE 4.1C: Selected statistical features for Figure 4.3 before and after despeckle filtering for increasing window size (from [3 × 3] to [25 × 25]) and for five iterations of the despeckle filter *DsFlsmv*

	WINDOW SIZE											
FEATURE	**3**	**5**	**7**	**9**	**11**	**13**	**15**	**17**	**19**	**21**	**23**	**25**
	36	35	36	36	36	36	36	36	36	36	36	36	35
Median	37	38	39	40	40	41	41	41	41	41	41	41	41
^{2}	21	18	18	17	17	17	17	17	17	16	16	16	16
^{3}	0.2	0.06	0.2	0.3	0.3	0.4	0.4	0.4	0.4	0.4	0.4	0.4	0.5
^{4}	2.8	2.4	2.1	2	2	2	2	2	2	1.9	1.9	1.9	1.9
Contrast	76	14	6	4	3	3	3	4	4	4	5	5	6
$C = (^2/)100$	58	51	50	47	47	47	47	47	47	44	44	44	46
CSR*100		0.44	0.44	0.44	0.44	0.44	0.44	0.44	0.44	0.44	0.44	0.44	0.46

Table 4.1d tabulates the statistical features, , median, ^{2}, ^{3}, ^{4}, the NGTDM contrast, the speckle index, C (3.13), and the contrast-speckle-ratio, CSR (3.14), for the artificial image, and the 11 filters illustrated in Figure 4.4. The filters are categorized in linear filtering, nonlinear filtering, diffusion filtering, and wavelet filtering, as introduced in Chapter 2. Also, the number of iterations for each despeckle filter is given, which was selected based on the speckle index, C, and on the visual perception of the two vascular experts. When C was minimally changing, then the filtering process was stopped. As shown in Table 4.1d, all filters reduced C with the exception of the *DsFhomo* filter, which exhibited the worst performance as it moves the mean of the image, , to a darker gray-level

---

FIGURE 4.3: Original phantom image given in (a), and the despeckled phantom images after the application of the *DsFlsmv* filter for increasing pixel moving window size from [3 × 3] to [25 × 25] given in (b)–(l). The number of iterations was for all cases five. (a) Original. (b) [3 × 3]. (c) [5 × 5]. (d) [7 × 7]. (e) [9 × 9]. (f) [11 × 11]. (g) [13 × 13]. (h) [15 × 15]. (i) [17 × 17]. (j) [19 × 19]. (k) [21 × 21]. (l) [23 × 23].

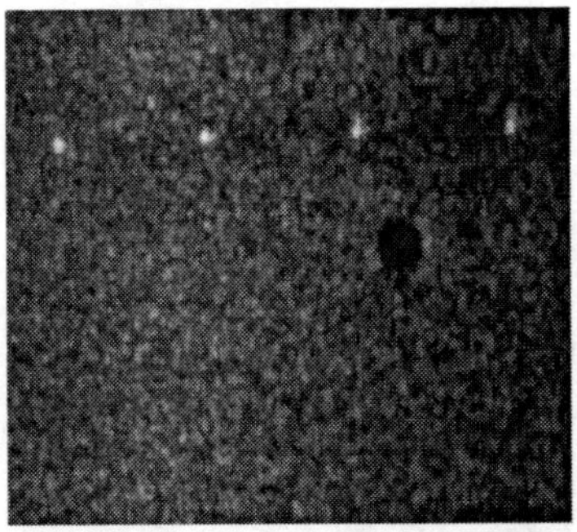
(a) Original

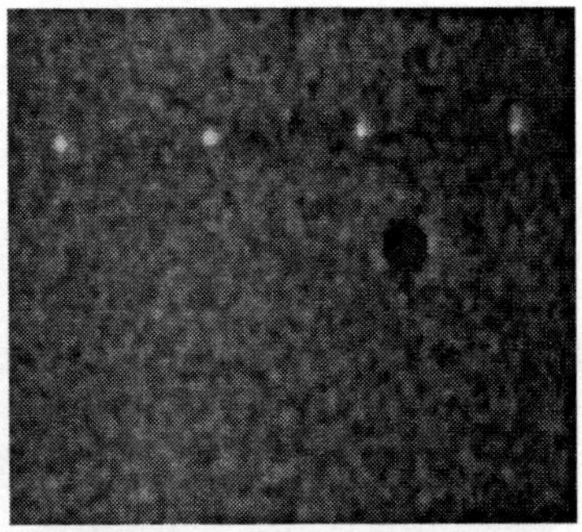
(b) [3x3]

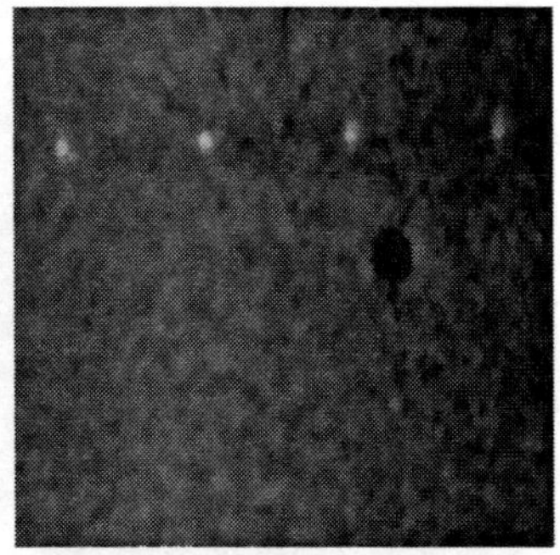
(c) [5x5]

(d) [7x7]

(e) [9x9]

(f) [11x11]

(g) [13x13]

(h) [15x15]

(i) [17x17]

(j) [19x19]

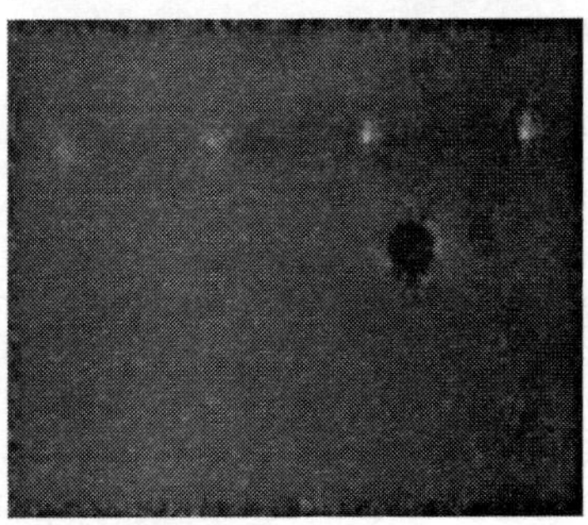
(k) [21x21]

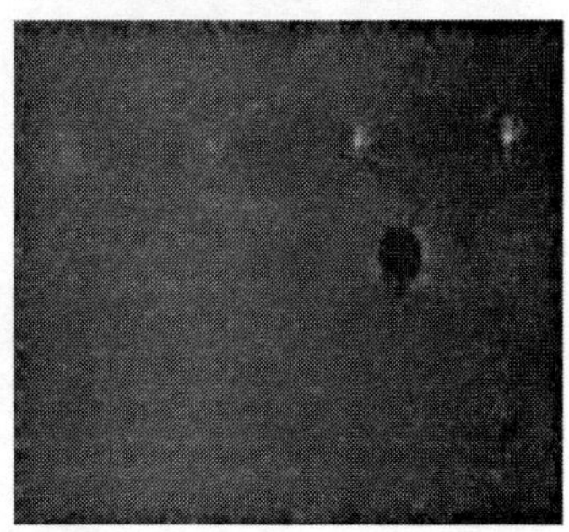
(l) [23x23]

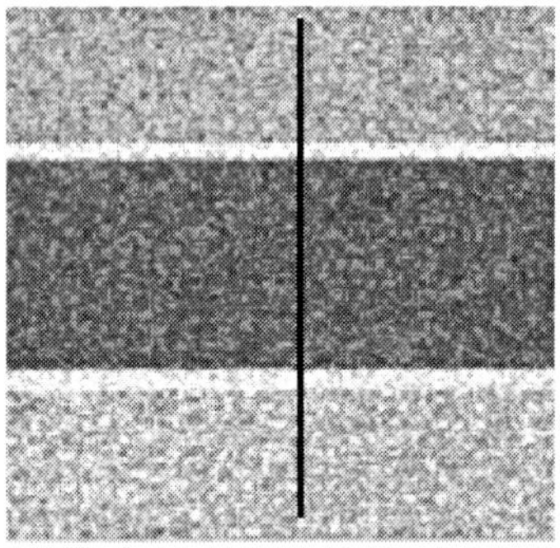

(a) original noisy image

(b) *DsFlsmv*

(c) *DsFlsminsc*

(d) *DsFwiener*

(e) *DsFmedian*

(f) *DsFls*

(g) *DsFhomog*

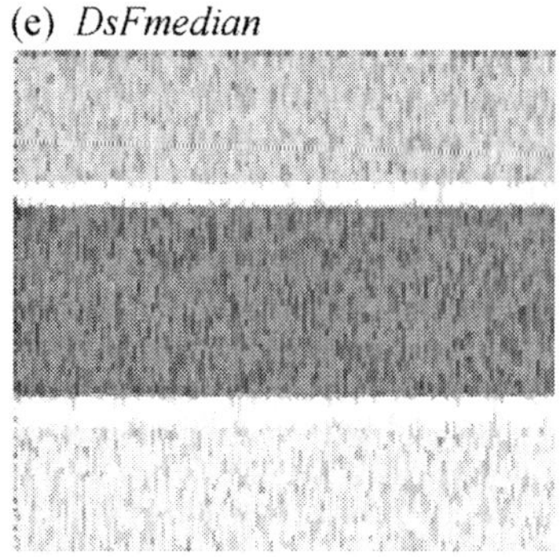

(h) *DsFgf4d*

(i) *DsFhomo*

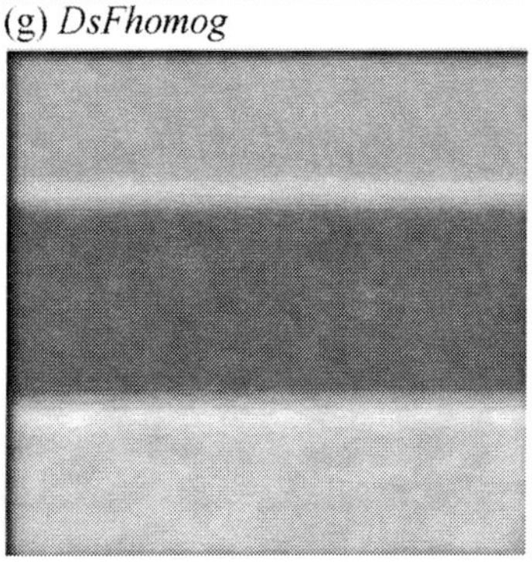

(j) *DsFad*

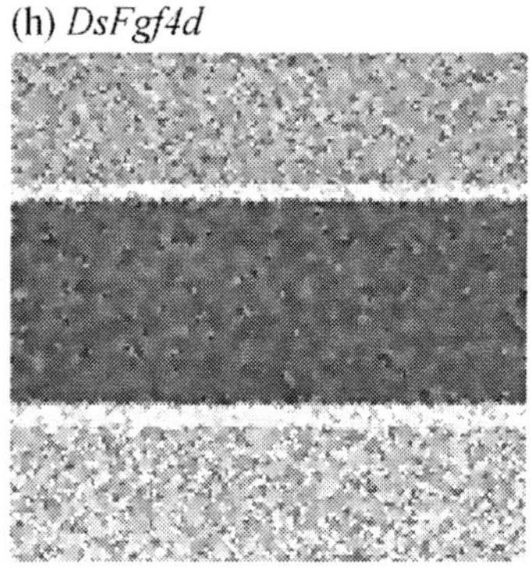

(k) *DsFnldif*

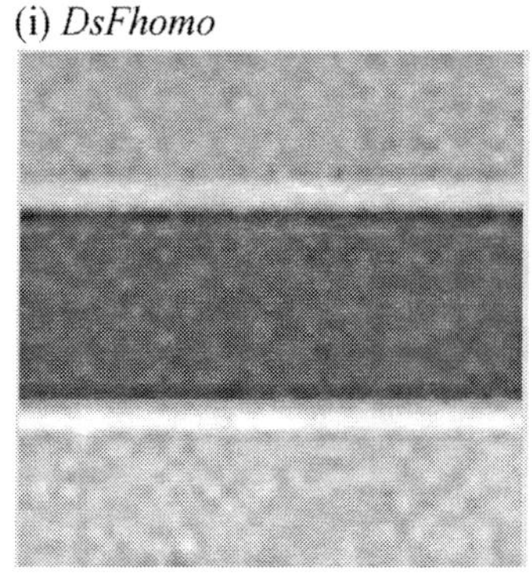

(l) *DsFwaveltc*

value, thus making the image darker. The CSR is better for the *DsFhomo*, *DsFgf4d*, *DsFlsminsc*, *DsFwaveltc*, *DsFwiener*, *DsFmedian*, and *DsFlsmv*. Filters that reduced the variance, ^{2}, while preserving the mean, , and the median compared to the original image, were: *DsFhomo*, *DsFls*, *DsFwiener*, *DsFwaveltc*, *DsFad*, *DsFhomog*, *DsFmedian*, and *DsFlsmv*. The contrast of the image is increased by the filters *DsFgf4d* (enormously), *DsFhomo*, *DsFlsminsc*, *DsFls*, *DsFmedian*, and *DsFhomog*, and it is decreased by the filters *DsFad*, *DsFwiener*, *DsFwaveltc*, and *DsFlsmv*. It is noted that filters *DsFgf4d*, *DsFlsmv*, and *DsFlsminsc* reduced *C*, increased CSR, *DsFlsmv* reduced the contrast, whereas *DsFlsminsc* increased the contrast.

4.1.3 Real Carotid Ultrasound Image

Figure 4.6 shows an ultrasound image of the carotid together with the despeckled images. The best visual results as assessed by the two experts were obtained for the filters *DsFlsmv* and *DsFlsminsc*, whereas the filters *DsFgf4d*, *DsFad*, and *DsFnldif* also showed good visual results but smoothed the image, loosing subtle details and affecting the edges. Filters that showed a blurring effect are the *DsFmedian*, *DsFwiener*, *DsFhomog*, and *DsFwaveltc*. Filters *DsFwiener*, *DsFhomog*, and *DsFwaveltc* showed poorer visual results. The number of iterations for each filter is given in Table 4.2.

4.1.4 Real Cardiac Ultrasound Images

Figure 4.7 shows two original ultrasound cardiac images in (a) and (e) (also shown in chapter 1, Figure 1.2) and the despeckled images in (b), (c), (d) and (f), (g), (h) with filters *DsFsrad*, *DsFlsmv*, and *DsFgf4d*, respectively. The moving sliding window applied to all images was [7 × 7] pixels. The best visual results as assessed by the two experts were obtained by the filters *DsFlsmv* (see Figure 4.7b and f) after four iterations, the filter *DsFsrad* (see Figure 4.7d and h) after 50 iterations and a coefficient of variation 0.025, and the *DsFgf4d* (see Figure 4.7c and g) after four iterations. The rest of the filters presented in Chapter 3 showed also good visual results but smoothed the image loosing subtle details, affecting also the edges.

FIGURE 4.4: Original noisy image of an artificial carotid artery given in (a), and the application of the 11 despeckle filters given in (b)–(l). (Vertical line given in (a) defines the position of the line intensity profiles plotted in Figure 4.5). (a) original noisy image. (b) *DsFlsmv*. (c) *DsFlsminsc*. (d) *DsFwiener*. (e) *DsFmedian*. (f) *DsFls*. (g) *DsFhomog*. (h) *DsFgf4d*. (i) *DsFhomo*. (j) *DsFad*. (k) *DsFnldif*. (l) *DsFwaveltc*.

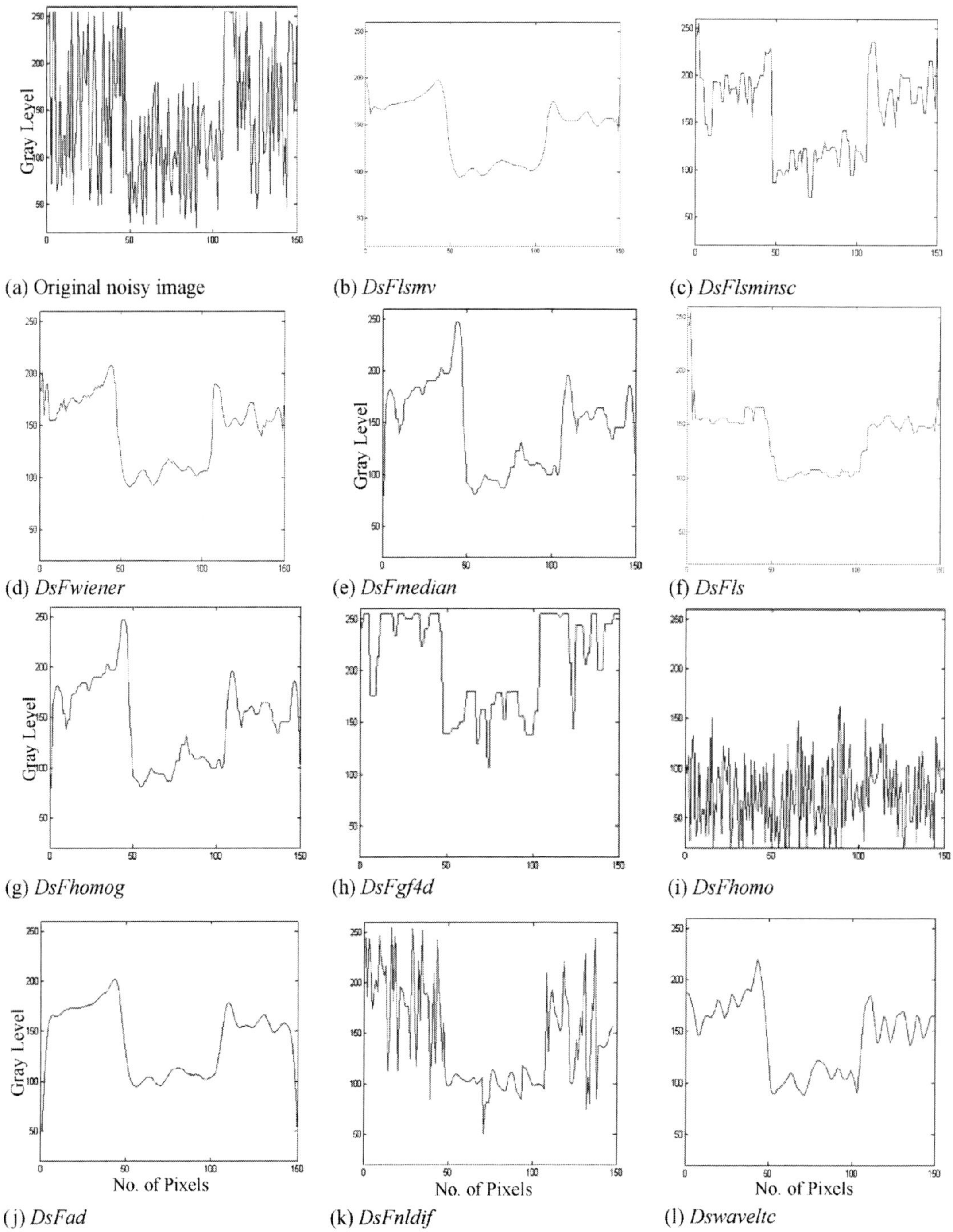

(a) Original noisy image (b) *DsFlsmv* (c) *DsFlsminsc*

(d) *DsFwiener* (e) *DsFmedian* (f) *DsFls*

(g) *DsFhomog* (h) *DsFgf4d* (i) *DsFhomo*

(j) *DsFad* (k) *DsFnldif* (l) *Dswaveltc*

4.2 EVALUATION OF DESPECKLE FILTERING ON CAROTID PLAQUE IMAGES BASED ON TEXTURE ANALYSIS

4.2.1 Distance Measures

Despeckle filtering and texture analyses were carried out on 440 ultrasound images of the carotid. Table 4.2 tabulates the results of $feat\_dis_z$(3.3) and *Score_Dis* (3.4) for SF, SGLDM range of values and NGTDM feature sets for the 10 despeckle filters. The results of these feature sets are presented only, as they were the ones with the best performance. The filters are categorized as linear filtering, nonlinear filtering, diffusion filtering, and wavelet filtering, as introduced in Chapter 2. Also, the number of iterations for each filter is given, which was selected based on *C* and on the visual evaluation of the two experts. When *C* was minimally changing, then the filtering process was stopped. The bold values represent the values that showed an improvement after despeckle filtering compared to the original. The last row in each subtable shows the *Score_Dis* for all features, where the highest value indicates the best filter in the subtable.

In addition, a total score distance *Score_Dis_T* was computed for all feature sets shown in the last row of Table 4.2. Some of the despeckle filters, shown in Table 4.2, are changing a number of texture features, by increasing the distance between the two classes (positive values in Table 4.2) and therefore making the identification and separation between asymptomatic and symptomatic plaques more feasible. A positive feature distance shows improvement after despeckle filtering, whereas a negative shows deterioration.

In the first part of Table 4.2, the results of the SF features are presented, where the best *Score_Dis* is given for the filter *DsFhomo* followed by the *DsFlsminsc*, *DsFlsmv*, *DsFhomog*, *DsFnldif*, *DsFwaveltc*, *DsFmedian*, and *DsFwiener*, with the worst *Score_Dis* given by *DsFgf4d*. All filters reduced the speckle index, *C*. Almost all filters reduced significantly the variance, ^{2}, and the kurtosis, ^{3}, of the histogram, as it may be seen from the bold values in the first part of Table 4.2.

In the second part of Table 4.2, the results of the SGLDM-Range of values features set are tabulated. The filters with the highest *Score_Dis* in the SGLDM range of values features set are *DsFhomo*, *DsFlsminsc*, *DsFmedian*, *DsFad*, and *DsFhomog*, whereas all the other filters (*DsFnldif*, *DsFwiener*, *DsFwaveltc*, *DsFgf4d*, *DsFlsmv)* are presenting a negative *Score_Dis*. Texture features,

FIGURE 4.5: Line profiles of the line illustrated in Figure 4.4a for the original noisy image (a), and the 11 despeckled images given in (b)–(l). (a) Original noisy image. (b) *DsFlsmv*. (c) *DsFlsminsc*. (d) *DsFwiener*. (e) *DsFmedian*. (f) *DsFls*. (g) *DsFhomog*. (h) *DsFgf4d*. (i) *DsFhomo*. (j) *DsFad*. (k) *DsFnldif*. (l) *DsFwaveltc*.

TABLE 4.1D: Selected statistical features for Figure 4.4 before and after despeckle filtering

FEATURE	ORIGINAL IMAGE	LINEAR FILTERING			
		DsFlsmv	*DsFlsminsc*	*DsFwiener*	*DsFmedian*
NO. OF ITERATIONS		4	1	2	2
	138	145	157	145	145
Median	132	151	162	157	152
2	53	41	46	37	40
3	0.85	-0.1	0.09	-0.2	0.07
4	2	2	1.8	1.6	2
Contrast	124	68	**239**	27	**141**
C = (2/)100	38	**28**	**29**	**26**	**28**
CSR*100		99	**263**	**101**	**100**

Bold values show improvement after despeckle filtering. Source [7], © IEEE 2005.

which improved in most of the filters, are the contrast, correlation, sum of squares variance, sum average, and sum variance.

In the third part of Table 4.2, for the NGTDM feature set, almost all filters showed an improvement in *Score_Dis* . The best filters in the NGTDM feature set were, the *DsFhomo*, *DsFlsminsc*, *DsFhomog*, and *DsFlsmv*. The texture features that improved at most were the completion, coarseness, and contrast. The completion of the image was increased by all filters.

Finally, in the last row of Table 4.2, the total score distance, *Score_Dis_T* , for all feature sets is shown, where best values were obtained by the filters *DsFhomo*, *DsFlsminsc*, *DsFmedian*, *DsFlsmv*, *DsFhomog*, and *DsFad*.

4.2.2 Univariate Statistical Analysis

Table 4.3 shows the results of the rank sum test, which was performed on the SGLDM range of values features set of Table 4.2, for the 10 despeckle filters. The test was performed to check if significant differences exist among the features computed on the 440 original and the 440 despeckled images. Filters that resulted with the most significant number of features after despeckle filtering as shown with the score row of Table 4.3 were the following: *DsFlsmv* (seven), *DsFgf4d* (six),

NONLINEAR FILTERING				DIFFUSION		WAVELET
DsFls	*DsFhomog*	*DsFgf4d*	*DsFhomo*	*DsFad*	*DsFnldif*	*DsFwaveltc*
3	3	3	2	20	5	5
143	145	176	55	139	143	146
157	156	157	55	152	132	156
33	40	46	24	39	51	38
-0.2	0.02	0.07	0.36	-0.35	0.44	-0.09
1.8	1.8	1.8	4	2	2	1.6
201	**132**	**1072**	**340**	**26**	60	50
23	**28**	**26**	44	**28**	**36**	**26**
74	**100**	**527**	**1305**	**14**	**68**	**115**

DsFlsminsc (five), and *DsFnldif* (four). The rest of the filters gave a lower number of significantly different features. Features that showed a significant difference after filtering were the inverse difference moment, IDM (eight), angular second moment, ASM (seven), sum of entropy (five), contrast (three), correlation (three), sum of squares variance (two), SOSV (two), and sum variance, Var (two).These features were mostly affected after despeckle filtering, and they were significantly different.

4.2.3 kNN Classifier

Table 4.4 shows the percentage of correct classifications score for the kNN classifier with $k = 7$ for classifying a subject as asymptomatic or symptomatic. The classifier was evaluated using the leave one out method [44] on 220 asymptomatic and 220 symptomatic images on the original and despeckled images. The percentage of correct classifications score is given for the following feature sets: statistical features, SF; spatial gray-level dependence matrix mean values, SGLDMm; spatial gray-level dependence matrix range of values, SGLDMr; gray-level difference statistics, GLDS; neighborhood gray-tone difference matrix, NGTDM; statistical feature matrix, SFM; laws texture energy measures, TEM; fractal dimension texture analysis, FDTA; and Fourier power spectrum,

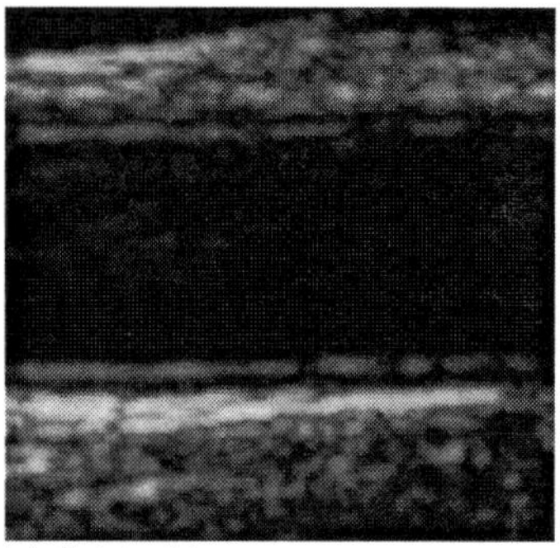

(a) *Original image*

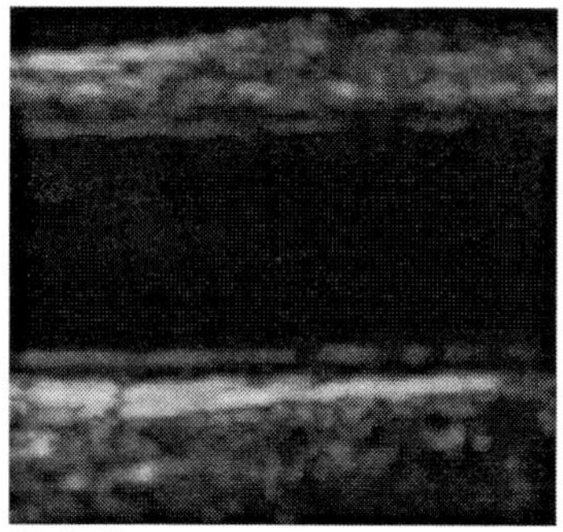

(b) *DsFlsmv*

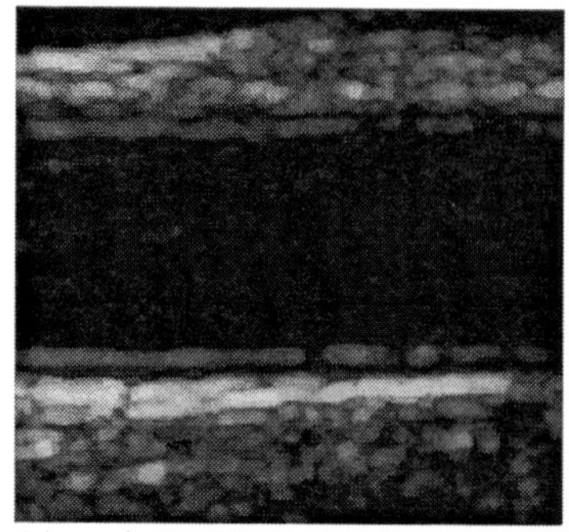

(c) *DsFlsminsc*

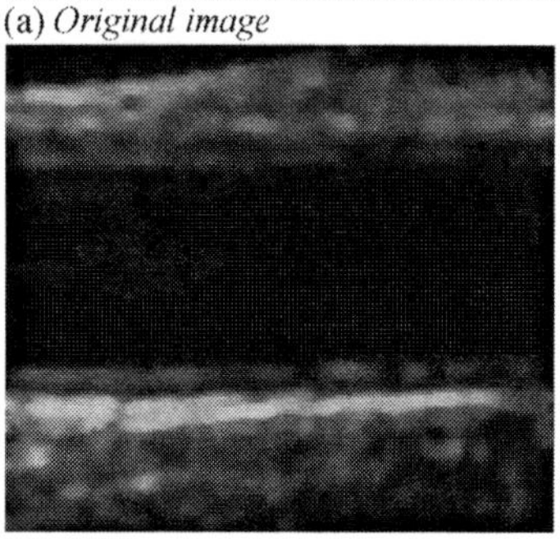

(d) *DsFwiener*

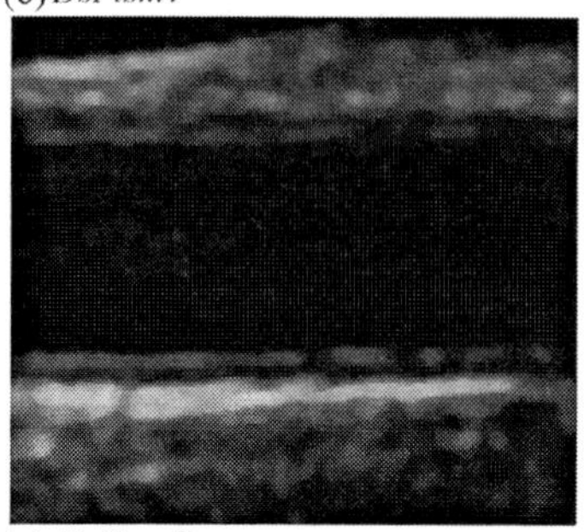

(e) *DsFmedian*

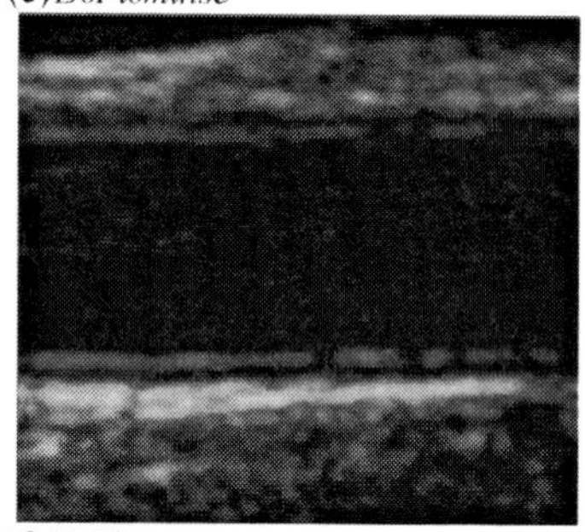

(f) *DsFhomog*

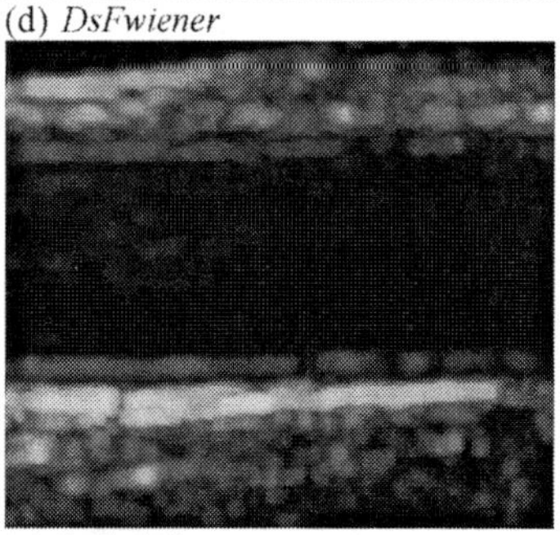

(g) *DsFgf4d*

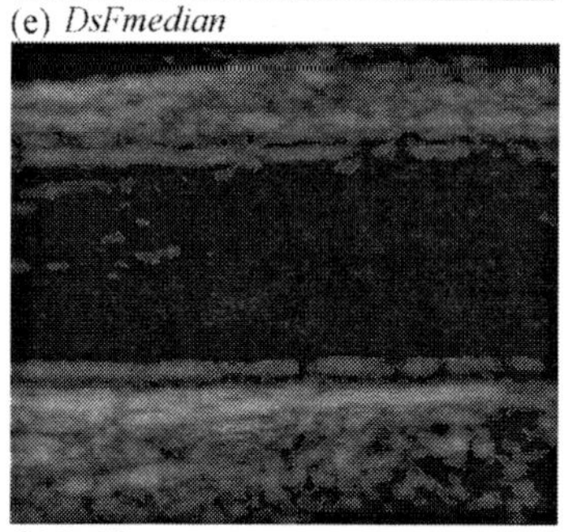

(h) *DsFhomo*

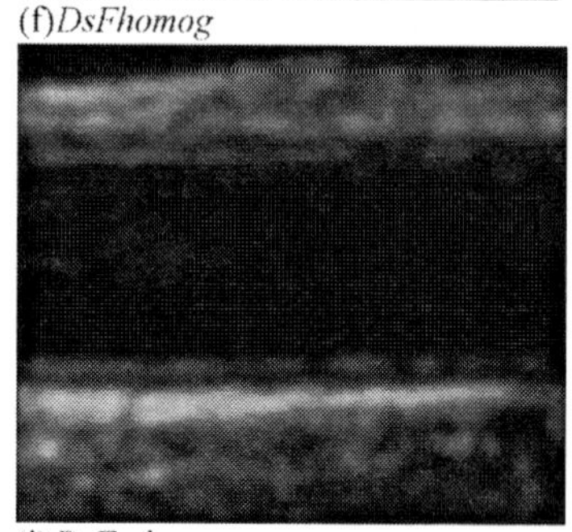

(i) *DsFad*

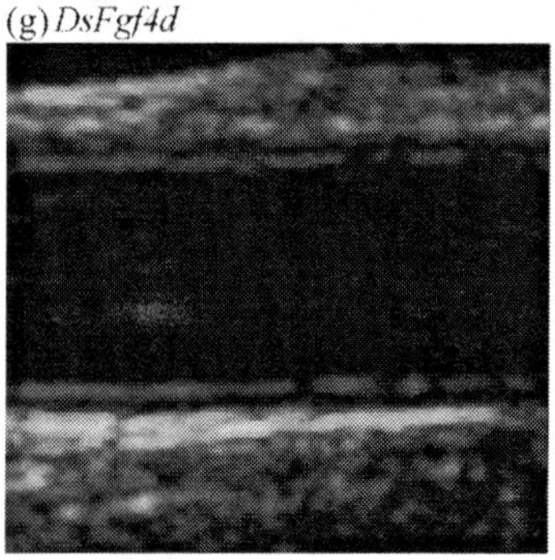

(j) *DsFnldif*

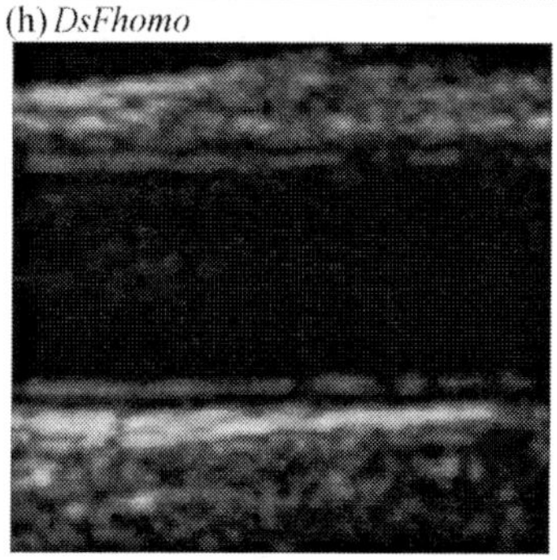

(k) *DsFwaveltc*

FPS. Filters that showed an improvement in classifications success score compared to that of the original image set were, in average (last row of Table 4.4), the filter *DsFhomo* (3%), *DsFgf4d* (1%), and *DsFlsminsc* (1%).

Feature sets, which benefited mostly by the despeckle filtering, were (last column in Table 4.4) the SF (seven), Tem (seven), SGLDm (four), GLDS (four), and NGTDM (four), when counting the number of cases that the correct classifications score was improved. Less improvement was observed, for the feature sets FDTA, SFM, FPS, and SGLDMr. For the feature set SGLDMr, better results are given for the *DsFlsminsc* filter with an improvement of 2%. This is the only filter that showed an improvement for this class of features. For the feature set TEM, the filter *DsFlsmv* shows the best improvement with 9%, whereas for the FPS feature set, the filter *DsFlsminsc* gave the best improvement with 5%. The filter *DsFlsminsc* showed improvement in the GLDS and NGTDM feature sets, whereas the filter *DsFlsmv* showed improvement for the feature sets SF and TEM.

4.3 IMAGE QUALITY AND VISUAL EVALUATION

Table 4.5 tabulates the image quality evaluation metrics presented in Section 3.9, for the 220 asymptomatic and 220 symptomatic ultrasound images between the original and the despeckled images, respectively. Best values were obtained for the *DsFnldif*, *DsFlsmv*, and *DsFwaveltc* with lower MSE, RMSE, Err3, and Err4 and higher SNR and PSNR. The GAE was 0.00 for all cases, and this can be attributed to the fact that the information between the original and the despeckled images remains unchanged. Best values for the universal quality index, *Q*, and the structural similarity index, SSIN, were obtained for the filters *DsFlsmv* and *DsFnldif*.

Table 4.6a shows the results of the visual evaluation of the original and despeckled images made by two experts, a cardiovascular surgeon and a neurovascular specialist. They evaluated 100 ultrasound images before and after despeckle filtering [50 asymptomatic (A) and 50 symptomatic (S)]. For each case, a total of 10 images were evaluated (one original and nine filtered). For each case, for each image, the experts assigned a score in the one to five scale based on subjective criteria. Therefore, the maximum score for a filter is 500, if the expert assigned the score of five for all the 100 images. For each filter, the score was divided by five to be expressed in percentage

---

FIGURE 4.6: Original ultrasound image of the carotid artery (2–3 cm proximal to bifurcation) given in (a), and the despeckled filtered images given in (b)–(k). (a) Original image. (b) *DsFlsmv*. (c) *DsFlsminsc*. (d) *DsFwiener*. (e) *DsFmedian*. (f) *DsFhomog*. (g) *DsFgf4d*. (h) *DsFhomo*. (i) *DsFad*. (j) *DsFnldif*. (k) *DsFwaveltc*.

TABLE 4.2: Feature distance [Eq. (3.3)] and *Score_Dis* [Eq. (3.4)] For SF, SGLDM range of values, and NGTDM texture feature sets between asymptomatic and symptomatic carotid plaque ultrasound images

	LINEAR FILTERING			
FEATURE	***DsFlsmv***	***DsFlsminsc***	***DsFwiener***	***DsFmedian***
NO. OF ITERATIONS	**4**	**1**	**2**	**2**
	Statistical features			
Mean	**14**	**22**	**19**	**4**
Median	5	17	26	5
^{2}	**18**	**38**	**18**	**7**
^{3}	**12**	**16**	**5**	**9**
^{4}	12	14	7	6
C	0.4	0.3	0.3	0.4
Score_dis	27	45	9	9
	SGLDM range of values—spatial gray-level dependence matrix			
ASM	21	0.5	29	2
Contrast	**47**	**107**	**14**	**64**
Correlation	**12**	**59**	**15**	**24**
SOSV	**9**	**40**	**18**	**10**
IDM	50	11	48	2
SAV	**17**	**24**	**23**	**7**
Var	**19**	**38**	**18**	**9**
Entr	34	14	49	**3**
Score_dis	1	243	38	121
	NGTDM—neighborhood gray-tone difference matrix			
Coarseness	**30**	**87**	**4**	**9**
Contrast	**7**	0.3	9	**8**
Busyness	**17**	**26**	30	**8**
Completion	**64**	**151**	**21**	**53**
Score_dis	118	264	14	78
Score_dis T	144	551	43	208

Bold values show improvement after despeckle filtering. ASM, angular second moment; SOSV, sum of squares variance; IDM, inverse difference moment; SAV, sum average; Var, sum variance. Source [7], © IEEE 2005.

NONLINEAR FILTERING			DIFFUSION		WAVELET
DsFhomog	*DsFgf4d*	*DsFhomo*	*DsFad*	*DsFnldif*	*DsFwaveltc*
1	3	1	20	5	5
11	**3**	**164**	**18**	**5**	**15**
5	15	**110**	29	6	15
13	2	**140**	**9**	**7**	**18**
7	0.1	**149**	**17**	**7**	**8**
4	3	**117**	21	**6**	9
0.3	0.4	0.08	0.3	0.4	0.3
22	17	680	6	19	17
4	8	47	25	17	20
32	3	**165**	**104**	**13**	**22**
5	**2**	**10**	**54**	4	4
16	2	**101**	**9**	**8**	**20**
29	8	**94**	54	34	43
15	**3**	**169**	**22**	**6**	**18**
15	2	**90**	**9**	**8**	**20**
19	4	11	47	30	36
21	22	571	72	50	23
16	7	**72**	36	37	33
0.4	4	**105**	**5**	27	15
1	4	**48**	14	39	8
80	**2**	**150**	**63**	**18**	**27**
66	13	375	18	85	13
108	52	1626	84	116	19

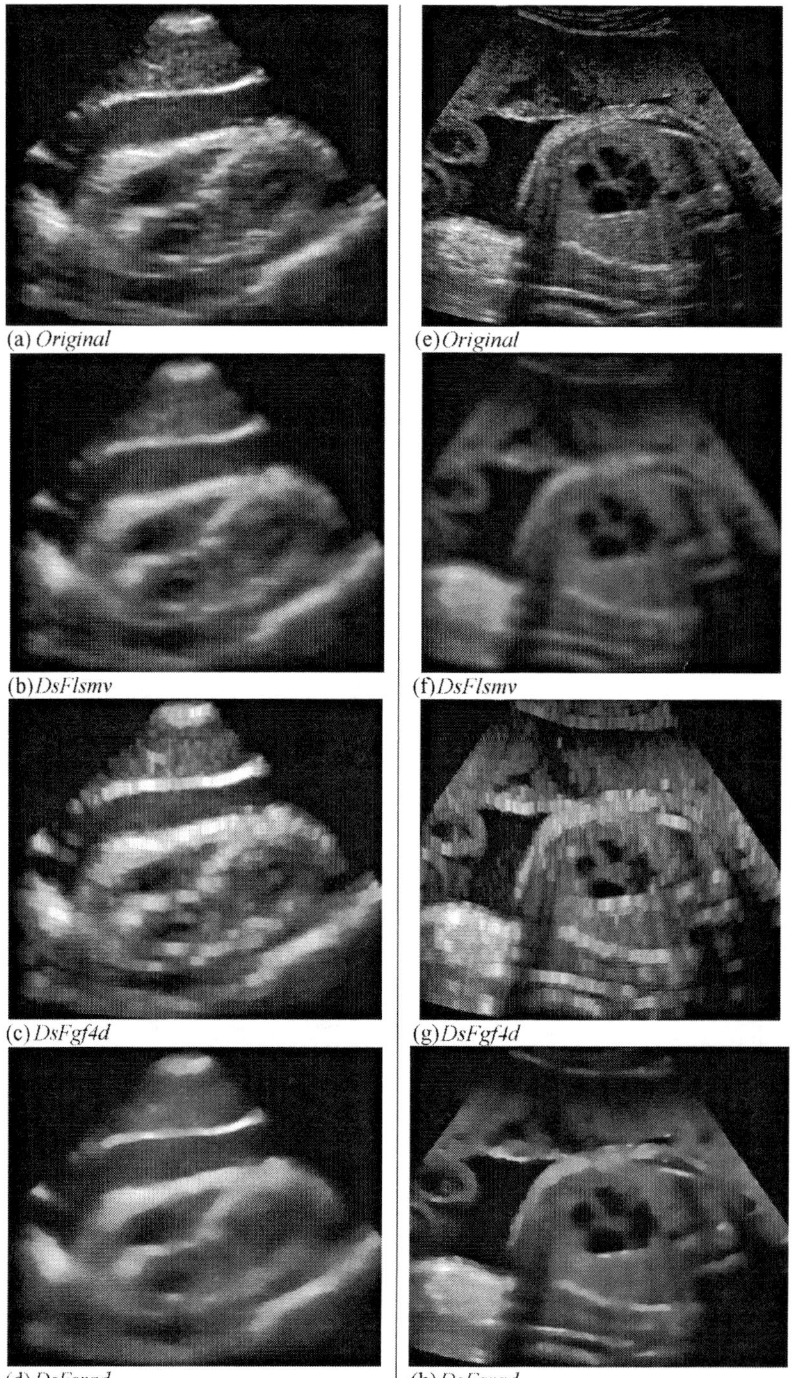

(a) *Original* (e) *Original*

(b) *DsFlsmv* (f) *DsFlsmv*

(c) *DsFgf4d* (g) *DsFgf4d*

(d) *DsFsrad* (h) *DsFsrad*

format. The last row of Table 4.6.a presents the overall average percentage (%) score assigned by both experts for each filter.

For the cardiovascular surgeon, the average score, showed that the best despeckle filter is the *DsFlsmv* with a score of 62%, followed by *DsFgf4d*, *DsFmedian*, *DsFhomog*, and *original* with scores of 52%, 50%, 45%, and 41% respectively. For the neurovascular specialist, the average score showed that the best filter is the *DsFgf4d* with a score of 72%, followed by *DsFlsmv*, *original*, *DsFlsminsc*, and *DsFmedian* with scores of 71%, 68%, 68% and 66%, respectively. The overall average % score shows that the highest score was given to the filter *DsFlsmv* (67%), followed by *DsFgf4d* (62%), *DsFmedian* (58%), and *original* (54%). It should be emphasized that the despeckle filter *DsFlsmv* is the only filter that was graded with a higher score than the original by both experts for the asymptomatic and symptomatic image sets.

We may observe a difference in the scorings between the two vascular specialists, and this is because the cardiovascular surgeon is primarily interested in the plaque composition and texture evaluation, whereas the neurovascular specialist is interested to evaluate the degree of stenosis and the lumen diameter to identify the plaque contour. Filters *DsFlsmv* and *DsFgf4d* were identified as the best despeckle filters, by both specialists as they improved visual perception with overall average scores of 67% and 62%, respectively. The filters *DsFwaveltc* and *DsFhomo* were scored by both specialists with the lowest overall average scores of 28% and 29%, respectively.

Table 4.6b shows the results of the visual perception evaluation made by the same experts, 1 year after the first visual evaluation. The visual perception evaluation was repeated to assess the intra-observer variability between the same experts and was performed under the same conditions as the first visual evaluation.

For the cardiovascular surgeon, the average score, showed that the best despeckle filter is again the *DsFlsmv* with a score of 61%, followed by *DsFmedian*, *DsFgf4d*, *DsFls*, *DsFhomog*, and *original* with scores of 60%, 52%, 49%, 40%, and 36%, respectively. For the neurovascular expert, the average score showed that the best filter is the *DsFlsminsc* with a score of 68%, followed by *DsFgf4d*, *DsFlsmv*, *original*, and *DsFmedian* with scores of 67%, 64%, 63%, and 61%, respectively. The overall average % score shows that the highest score was given to the filter *DsFlsmv* (63%), followed by

---

FIGURE 4.7: (a) and (e) Original cardiac ultrasound images (see also Chapter 1, Figure 1.2) and the despeckled images with the filters *DsFlsmv*, *DsFgf4d*, and *DsFsrad*, given in the left column and the right column for the first and second images, respectively. (a) Original. (b) *DsFlsmv*. (c) *DsFgf4d*. (d) *DsFsrad*. (e) Original. (f) *DsFlsmv*. (g) *DsFgf4d*. (h) *DsFsrad*.

TABLE 4.3: Wilcoxon rank sum test for the SGLDM range of values texture features applied on the 440 ultrasound images of carotid plaque before and after despeckle filtering

FEATURE	LINEAR FILTERING			NONLINEAR FILTERING	
	DsFlsmv	*DsFlsminsc*	*DsFwiener*	*DsFmedian*	*DsFhomog*
ASM	S	S	NS	NS	S
Contrast	S	NS	NS	NS	NS
Correlation	S	S	NS	NS	NS
SOSV	S	NS	NS	NS	NS
IDM	S	S	NS	S	S
SAV	NS	NS	NS	NS	NS
ΣVar	S	S	NS	NS	NS
ΣEntropy	S	S	NS	NS	NS
Score	7	5	0	1	2

The test shows with S significant difference after filtering at $p < 0.05$ and NS no significant difference after filtering at $p > 0.05$. ASM, angular second moment; SOSV, sum of squares variance; IDM, inverse difference moment; SAV, sum average; Var, sum variance; Score, illustrates the number of S. Source [7], © IEEE 2005.

DsFmedian (61%), *DsFgf4d* (60%), *DsFls* (54%), and *original* (50%). The intra-observer variability results in Table 4.6b shows a consistency in almost all results, with only very small differences among filters. The despeckle filter *DsFlsmv* is again, the only filter that was graded with a higher score than the original by both vascular experts for the asymptomatic and symptomatic images.

Both experts were in agreement that the best despeckle filters for visual perception are the *DsFlsmv*, *DsFlsminsc*, *DsFgf4d*, and *DsFmedian*, whereas the worst filters were the *DsFwaveltc* followed by the *DsFhomo* and *DsFnldif* (see also Table 4.6a and Table 4.6b). Furthermore, both experts agreed that almost all despeckle filters reduced the noise substantially, and images may be better visualized after despeckle filtering. By examining the visual results of Figures 4.1–4.7, the statistical results of Tables 4.1–4.5, and the visual evaluation of Table 4.6, we can conclude that the best filters are *DsFlsmv* and *DsFgf4d*, which may be used for both plaque composition enhancement and plaque texture analysis, whereas the filters *DsFlsmv*, *DsFgf4d*, and *DsFlsminsc* are more appropriate to identify the degree of stenosis and, therefore, may be used when the primary interest is to outline the plaque borders.

		DIFFUSION		WAVELET	SCORE
DsFgf4d	*DsFhomo*	*DsFad*	*DsFnldif*	*DsFwaveltc*	
S	NS	S	S	S	7
S	NS	NS	S	NS	3
S	NS	NS	NS	NS	3
S	NS	NS	NS	NS	2
S	S	NS	S	S	8
NS	NS	NS	NS	NS	0
NS	NS	NS	NS	NS	2
S	NS	NS	S	S	5
6	1	1	4	3	

4.4 SEGMENTATION OF THE INTIMA–MEDIA COMPLEX AND PLAQUE IN THE CCA BASED ON DESPECKLE FILTERING

Ultrasound measurements of the human carotid artery walls are conventionally obtained by manually tracing interfaces between tissue layers. In this section, we present results from a snakes segmentation technique [54] for detecting the intima–media layer of the far wall of the common carotid artery (CCA) in longitudinal ultrasound images (see Figure 4.8), by applying snakes, after normalization, speckle reduction, and normalization and speckle reduction. The intima–media thickness (IMT; see Figure 4.8) of the common carotid artery (CCA) can serve as an early indicator of the development of cardiovascular disease, like myocardial infarction and stroke [111]. Previous studies indicated that increase in the IMT of the CCA is directly associated with an increased risk of myocardial infarction and stroke, especially in elderly adults without any history of cardiovascular disease [111, 112]. Importantly, increased IMT was demonstrated to have a strong correlation with the presence of atherosclerosis elsewhere in the body and may thus be used as a descriptive index of individual atherosclerosis [113]. As vascular disease develops, local changes occur in arterial

TABLE 4.4: Percentage of correct classifications score for the kNN classifier with $k = 7$ for the original and the filtered image sets

FEATURE SET	NO. FEATURES	ORIGINAL	LINER FILTERING			
			DsFlsmv	*DsFlsminsc*	*DsFwiener*	*DsFmedian*
SF	5	59	**62**	**61**	**61**	57
SGLDMm	13	65	63	64	62	63
SGLDMr	13	70	66	**72**	64	66
GLDS	4	64	63	**66**	61	**69**
NGTDM	5	64	63	**68**	60	**69**
SFM	4	62	62	60	62	58
TEM	6	59	**68**	52	**60**	59
FDTA	4	64	63	**66**	53	**68**
FPS	2	59	54	**64**	59	58
Average		63	63	**64**	60	63

Bold values indicate improvement after despeckling. SF, statistical features; SGLDMm, spatial gray-level dependence matrix mean values; SGLDMr, spatial gray-level dependence matrix range of values; GLDS, gray-level difference statistics; NGTDM, neighborhood gray-tone difference matrix; SFM, statistical feature matrix; TEM, laws texture energy measures; FDTA, fractal dimension texture analysis; FPS, Fourier power spectrum. Source [7], © IEEE 2005.

structure, which thicken the innermost vessel layers known as intima–media complex (IMC). As disease progresses, the IMT initially increases diffusely along the artery and then becomes more focal, forming discrete lesions or plaques, which gradually grow and obstruct blood flow. Furthermore, these plaques can become unstable and rupture with debris transported distally by blood to obstruct more distal vessels. This is particular so if plaques develop internal pools of lipid covered only by a thin fibrous cap [113]. It is therefore important to accurately estimate the IMT.

The use of ultrasound provides a noninvasive method for estimating the IMT of human carotid arteries and is specially suited to dynamic analysis owing to its ability to deliver real-time video sequences. A B-mode ultrasound image shown in Figure 4.8 shows the IMC at the far wall of the carotid artery (echo zones Z5–Z6) as a pair of parallel bands, an echodense and an echolucent. The band Z5 and the leading edge of the band Z7 (adventitia), denoted as I5 and I7, define the far-wall IMT. With this understanding, the determination of the IMT at the far wall of the artery becomes equivalent to accurately detecting the leading echo boundaries I5 and I7. The lumen–intima and

NONLINEAR FILTERING			DIFFUSION		WAVELET	SCORE
DsFhomog	*DsFgf4d*	*DsFhomo*	*DsFad*	*DsFnldif*	*DsFwaveltc*	
63	59	**65**	**60**	52	**61**	7
69	**67**	**68**	61	**66**	63	4
65	70	69	64	65	65	1
64	**66**	**72**	59	58	62	4
63	**65**	57	60	61	62	3
55	**65**	**68**	59	56	55	2
66	**60**	**65**	53	**60**	**60**	7
53	62	**73**	55	54	62	3
59	59	59	52	48	55	1
62	**64**	**66**	58	58	61	

media–adventitia intensity interface of the far wall of the carotid artery is preferred for IMT measurements [54, 128]. It has been shown that the definition of the IMT, as shown in Figure 4.8, corresponds to the actual histological IMT [1, 54, 111].

Traditionally, the IMT is measured by manual delineation of the intima and the adventitia layer [1, 54, 112, 128]. Manual tracing of the lumen diameter (see Figure 4.8 Z4) and the IMT (see Figure 1 I5, I7) by human experts requires substantial experience; it is time consuming and varies according to the training, experience, and the subjective judgment of the experts. The manual measurements suffer therefore from considerable inter- and intra-observer variability [54, 112, 128].

Carotid artery atherosclerosis is the primary cause of stroke and the third leading cause of death in the United States. Almost twice as many people die from cardiovascular disease than from all forms of cancer combined. Atherosclerosis is a disease of the large- and medium-sized arteries, and it is characterized by plaque formation due to progressive intimal accumulation of lipid, protein,

TABLE 4.5: Image quality evaluation metrics computed for the 220 asymptomatic and 220 symptomatic images

FEATURE SET	LINEAR FILTERING			
	DsFlsmv	*DsFlsminsc*	*DsFwiener*	*DsFmedian*
	Asymptomatic images			
MSE	13	86	19	131
RMSE	3	9	4	10
Err3	7	17	5	25
Err4	11	26	7	41
GAE	0	0	0	0
SNR	25	17	23	16
PSNR	39	29	36	29
Q	0.83	0.78	0.74	0.84
SSIN	0.97	0.88	0.92	0.94
	Symptomatic images			
MSE	33	374	44	169
RMSE	5	19	6	13
Err3	10	33	9	25
Err4	16	47	11	38
GAE	0	0	0	0
SNR	24	13	22	16
PSNR	34	23	33	26
Q	0.82	0.77	0.7	0.79
SSIN	0.97	0.85	0.89	0.81

MSE, mean square error; RMSE, randomized mean square error; Err3 and Err4, Minowski metrics; GAE, geometric average error; SNR, signal-to-noise radio; PSNR, peak signal-to-noise radio; *Q*, universal quality index; SSIN, structural similarity index. Source [7], © IEEE 2005.

NONLINEAR FILTERING			DIFFUSION		WAVELET
DsFhomog	*DsFgf4d*	*DsFhomo*	*DsFad*	*DsFnldif*	*DsFwaveltc*
42	182	758	132	8	11
6	13	27	11	2	3
14	25	38	21	5	4
24	40	49	32	10	5
0	0	0	0	0	0
21	14	5	14	28	25
34	27	20	28	41	39
0.92	0.77	0.28	0.68	0.93	0.65
0.97	0.88	0.43	0.87	0.97	0.9
110	557	1452	374	8	23
10	23	37	19	3	5
20	43	51	31	5	6
30	63	64	43	7	8
0	0	0	0	0	0
17	12	5	12	29	25
28	21	17	23	39	36
0.87	0.75	0.24	0.63	0.87	0.49
0.94	0.85	0.28	0.81	0.97	0.87

TABLE 4.6A: Percentage scoring of visual evaluation of the original and despeckled images [50 asymptomatic (A) and 50 symptomatic (S)] by the experts

EXPERTS	A/S	ORIGINAL	LINEAR FILTERING		
			DsFlsmv	*DsFlsminsc*	*DsFmedian*
Cardiovascular surgeon	A	33	75	33	43
	S	48	49	18	57
average %		41	62	26	50
Neurovascular specialist	A	70	76	73	74
	S	66	67	63	58
average %		68	71	68	66
Overall average %		54	67	47	58

TABLE 4.6B: Percentage scoring of visual evaluation of the original and despeckled images [50 asymptomatic (A) and 50 symptomatic (S)] by the experts 1 year after the first visual evaluation

EXPERTS	A/S	ORIGINAL	LINEAR FILTERING			
			DsFlsmv	*DsFlsminsc*	*DsFmedian*	*DsFls*
Cardiovascular surgeon	A	28	57	43	62	49
	S	44	65	24	57	49
average %		36	61	34	60	49
Neurovascular expert	A	62	65	64	69	67
	S	64	62	71	53	51
average %		63	64	68	61	59
Overall average %		50	63	51	61	54

NONLINEAR FILTERING			DIFFUSION	WAVELET
DsFhomog	*DsFgf4d*	*DsFhomo*	*DsFnldif*	*DsFwaveltc*
47	61	19	43	32
43	42	20	33	22
45	52	19	38	27
63	79	23	52	29
45	65	55	41	28
54	72	39	47	28
50	62	29	43	28

NONLINEAR FILTERING			DIFFUSION	WAVELET
DsFhomog	*DsFgf4d*	*DsFhomo*	*DsFnldif*	*DsFwaveltc*
41	53	16	39	31
39	51	23	37	21
40	52	20	38	26
51	65	19	49	24
49	69	49	44	26
50	67	34	47	25
45	60	27	43	26

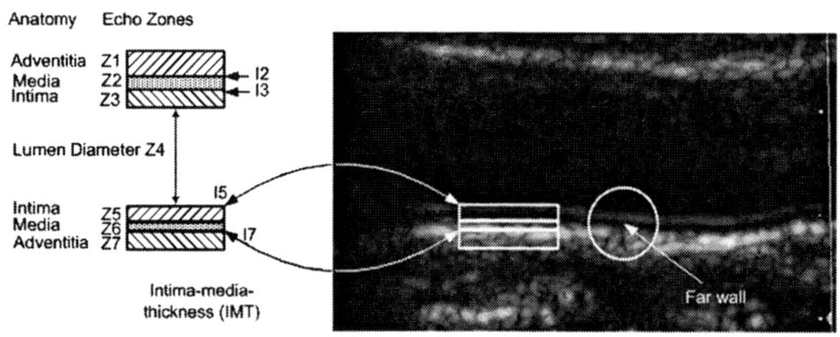

FIGURE 4.8: Illustration of the intima–media complex (IMC) of the far wall of the common carotid artery. The IMC consists of the intima band (Z5), the media band (Z6), and the far wall adventitia band (Z7). The IMT complex is defined as the distance between the blood intima interface line and the media adventitia interface line. Source [54], © MBEC 2007.

and cholesterol esters in the blood vessel wall [114], which reduces blood flow significantly. The risk of stroke increases with the severity of carotid stenosis and is reduced after carotid endarterectomy [115]. The degree of internal carotid stenosis is the only well-established measurement that is used to assess the risk of stroke [116]. Indeed, it is the only criterion at present used to decide whether carotid endarterectomy is indicated or not [94].

4.4.1 Intima–Media Complex and Plaque Segmentation

In a recent study [54], we developed and evaluated a snakes segmentation method for detecting the IMC in ultrasound imaging of the carotid artery after normalization and speckle-reduction filtering. It was shown that the application of normalization and speckle-reduction filtering before segmentation, improves both the manual and the automated IMC segmentation results. It should be noted that the despeckle filter *DsFlsmv*, with a moving sliding window of [5 × 5], was iteratively applied four times, on the area around the IMC and not to the whole image, on all images before segmentation of the IMC segmentation.

Figure 4.9 shows a longitudinal ultrasound image of the CCA with the manual delineations from the two experts (Figure 4.9b, Figure 4.9c), the automatic initial contour estimation (Figure 4.9d), and the Williams and Shah snakes segmentation results for the cases of no preprocessing (NP; Figure 4.9e), despeckled (DS; Figure 4.9f), normalized (N; Figure 4.9g), and normalized despeckled (NDS; Figure 4.9h). The detected IMT_{mean}, IMT_{max}, and IMT_{min} values, are shown with a double, single, and dashed line boxes, respectively. The results in Figure 4.9 showed that the

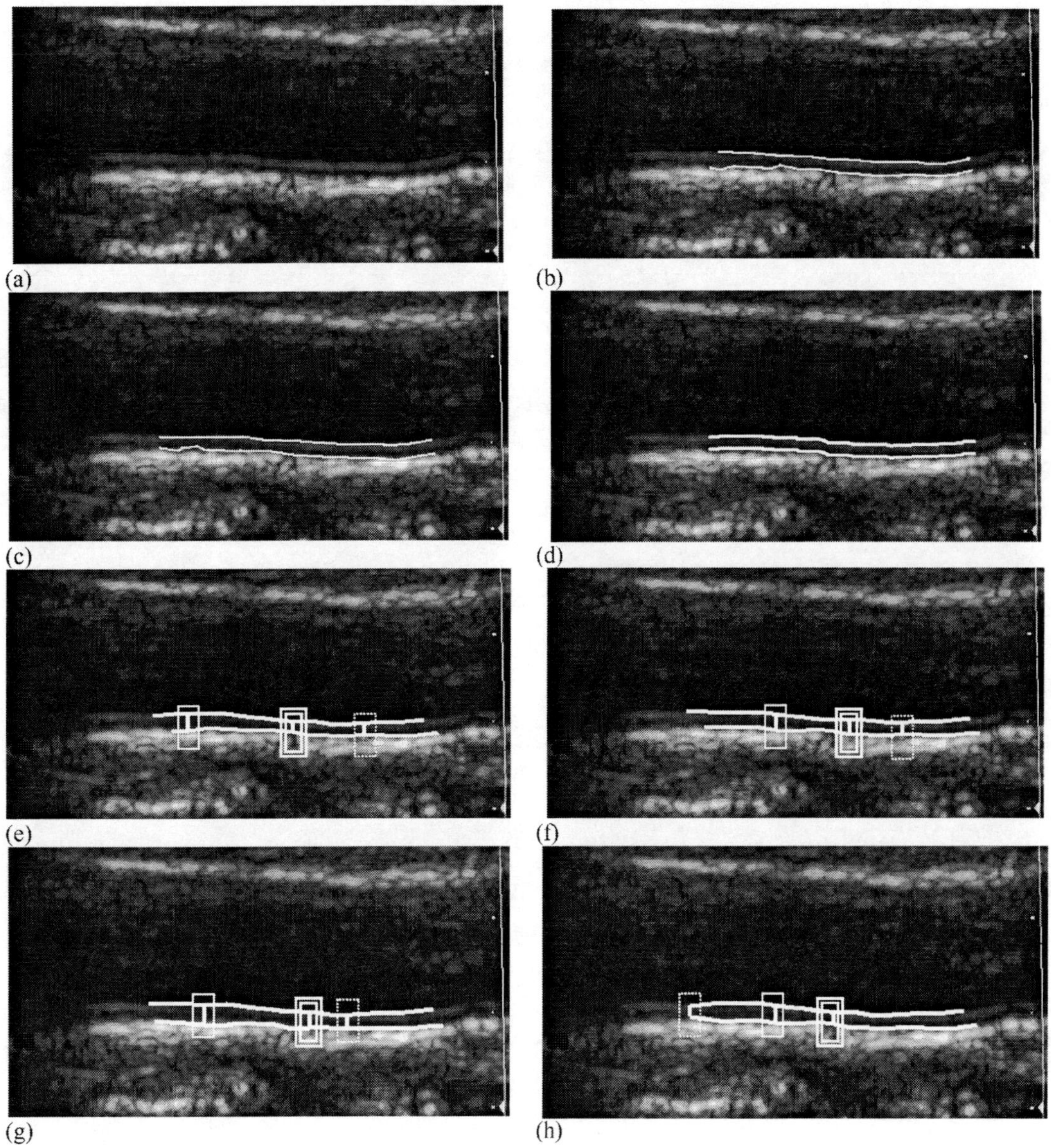

FIGURE 4.9: (a) Original longitudinal ultrasound image of the carotid artery, (b) manual delineation from Expert 1, (c) manual delineation from Expert 2, (d) initial contour estimation, and the segmentation results of the IMT for (e) no preprocessing (NP), (f) despeckled (DS), (g) normalized (N), and (h) normalized despeckled (NDS) images. The detected IMT_{mean}, IMT_{max}, and IMT_{min} are shown with a double, single, and dashed line boxes, respectively. Source [54], © MBEC 2007.

TABLE 4.7: Comparison between the manual and the snakes segmentation measurements for cases b–h in Figure 4.9

PARAMETER	MANUAL MEASUREMENTS				SNAKES SEGMENTATION MEASUREMENTS			
	EXPERT 1		EXPERT 2					
	M	MN	M	MN	NF	DS	N	NDS
IMT_{mean} (sd)	0.74 (0.14)	0.92 (0.11)	0.82 (0.12)	0.98 (0.15)	0.82 (0.22)	0.81 (0.21)	0.82 (0.19)	0.82 (0.18)
IMT_{min}	0.38	0.76	0.71	0.72	0.61	0.60	0.60	0.60
IMT_{max}	0.95	1.05	0.94	1.10	1.09	1.08	1.08	1.08
IMT_{median}	0.66	0.90	0.85	0.95	0.79	0.78	0.78	0.78

Measurements are in millimeters (mm). M, manual; MN, manual normalized; NF, no filtering; DS, despeckle; N, normalized; NDS, Normalized despeckled; sd, standard deviation. Source [54], © MBEC 2007.

IMT was detected in all snakes segmentation measurements but with variations between experts and methods.

The manual (for two different experts, Expert 1, Expert 2) and the automated IMT_{mean}, IMT_{min}, IMT_{max}, and IMT_{median} measurements for Figure 4.9 are presented in Table 4.7. The manual measurements are given for each expert (Expert 1 and Expert 2), in cases when manual measurements were carried out, without normalization (M) and with normalization (MN). The Williams and Shah snakes segmentation [105] measurements are given for the NF, DS, N, and NDS cases, and were in most of the cases, higher than the manual measurements, except in the MN case for both experts. The observed standard deviation, sd, values for the IMT_{mean}, was for the first expert, M (0.14), MN (0.11), for the second expert, M (0.12), MN (0.15), and for the snakes segmentation, NF (0.22), DS (0.21), N (0.19), and NDS (0.18), respectively. The results in Figure 4.9 and Table 4.7 show that the IMT was detected well in all snakes segmentation measurements but with varia-

FIGURE 4.10: Plaque segmentation results on a longitudinal ultrasound B-mode image of the carotid artery: (a) manual, (b) Williams and Shah, (c) Balloon, (d), Lai and Chin, (e) GVF snake, and (f) superimposition of segmentation contours computed in (b)–(e). [60], © IEEE 2007.

(a) Manual delineation.

(b) Williams & Shah.

(c) Balloon.

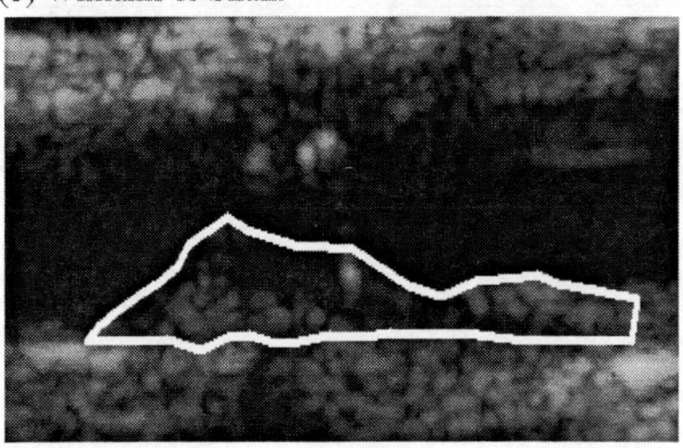

(d) Lai & Chin.

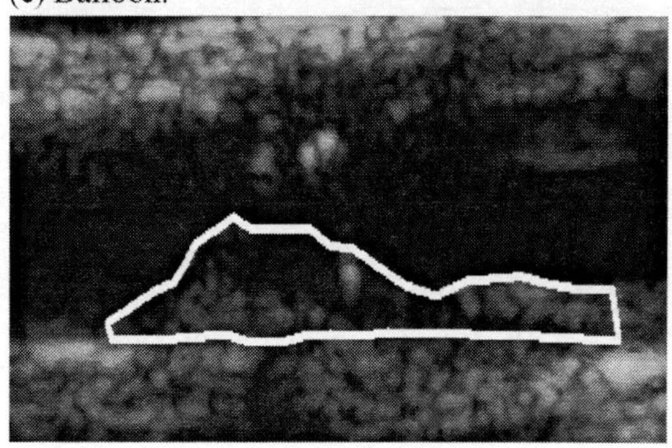

(e) GVF.

(f) Superimposition of segmentation contours computed in (b)-(e).

tions between experts and methods. The best visual results, as assessed by the two vascular experts, were obtained on the NDS, followed by N and DS images.

In another study [60], we proposed and evaluated an integrated plaque segmentation system based on normalization, speckle reduction filtering, and snakes segmentation. Four different snakes segmentation methods were investigated in Ref. [60], namely: (i) the Williams and Shah [105], (ii) the Balloon [106], (iii) the Lai and Chin [107], and (iv) the GVF [108]. These were applied on 80 plaque ultrasound images of the CCA. The comparison of the four different plaque snakes segmentation methods showed that the Lai and Chin segmentation method gave slightly better results, although these results were not statistically significant when compared with the other three snakes segmentation methods. It was also shown that the application of normalization and speckle-reduction filtering before segmentation, improves both the manual and the automated plaque segmentation results.

Figure 4.10 illustrates an original longitudinal ultrasound B-mode image of a carotid plaque with a manual delineation made by the expert in (a), and the results of the William and Shah segmentation in (b), the Balloon segmentation in (c), the Lai and Chin segmentation in (d), and the

TABLE 4.8: ROC analysis for the four different plaque segmentation methods and the manual delineations made by an expert on 80 ultrasound images of the carotid artery

SEGMENTATION METHOD	SYSTEM DETECTS	EXPERT DETECTS NO PLAQUE	EXPERT DETECTS PLAQUE	KI	OVERLAP INDEX
Williams and Shah	No plaque	TNF = 77.59%	FNF = 19.64%	78.86 %	67.60 %
	Plaque	FPF = 6.50%	TPF = 81.76%		
Balloon	No plaque	TNF = 77.12%	**FNF = 13.90%**	77.87 %	67.79 %
	Plaque	**FPF = 5.40%**	TPF = 80.35%		
Lai and Chin	No plaque	**TNF = 80.89%**	FNF = 15.59%	**80.66 %**	**69.30 %**
	Plaque	FPF = 5.86%	**TPF = 82.70%**		
GVF	No plaque	TNF = 79.44%	FNF = 14.90%	77.25 %	66.60 %
	Plaque	FPF = 6.30%	TPF = 79.57%		

Source [60], © IEEE 2005.

GVF segmentation in (e). Figure 4.10f shows the superimposition of the segmentation contours computed in Figure 4.10b–e. As illustrated in Figure 4.10f, both the manual and the snakes segmentation contours are visually very similar. It should be noted that the despeckle filter *DsFlsmv*, with a moving sliding window of [5 × 5], was iteratively four times applied on all images before segmentation.

Table 4.8 presents a comparison of the four different plaque snakes segmentation methods (Williams and Shah, Balloon, Lai and Chin, and GVF) with the manual segmentation as performed by an expert on 80 longitudinal ultrasound images of the carotid plaque as described in [60]. Although all methods demonstrated similar performance, the best overall performance was demonstrated by the Lai and Chin snakes segmentation method. The results showed that the Lai and Chin snakes segmentation method agrees with the expert in 80.89% of the cases, TNF, by correctly detecting no plaque, in 82.70% of the cases, TPF, by correctly detecting a plaque, disagrees with the expert in 15.59% of the cases, FNF, by detecting no plaque, and in 5.86% of the cases, FPF, by detecting a plaque. The similarity kappa index, KI, and the overlap index, for the Lai and Chin snakes segmentation method were the highest, equal to 80.66% and 69.3%, respectively.

The best FPF, and FNF, fractions were given by the Balloon snakes segmentation method, with 5.4% and 13.90%, respectively. The GVF snakes segmentation method, showed for this experiment the worst results with the lowest similarity kappa index, KI, (77.25%), and the lowest overlap index (66.6%).

TABLE 4.9: ROC analysis for the four different plaque segmentation methods and the manual delineations made by an expert on 80 ultrasound images of the carotid artery based on the sensitivity, *R*, specificity, Sp, precision, *P*, and 1-effectiveness measure, 1-*E*

SEGMENTATION METHOD	SENSITIVITY (*R*)	SPECIFICITY (SP)	PRECISION (*P*)	*F*=1–*E*
Williams and Shah	0.8176	0.9350	0.9263	0.8621
Balloon	0.8053	**0.9460**	0.9271	**0.8882**
Lai and Chin	**0.8270**	0.9416	**0.9338**	0.8851
GVF	0.7957	0.9370	0.9266	0.8824

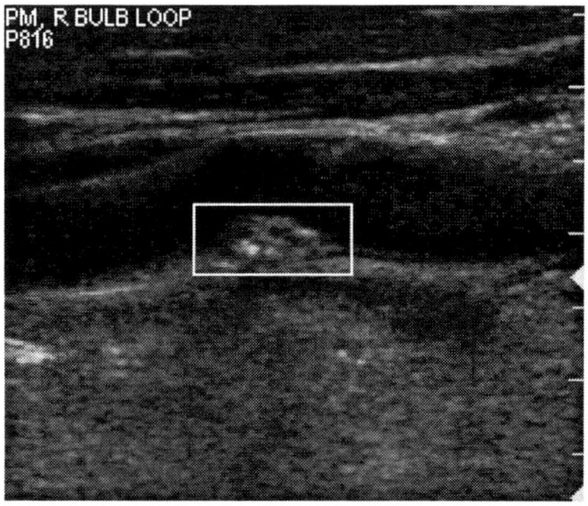

(a) Original Frame 1 (C=0. 863)

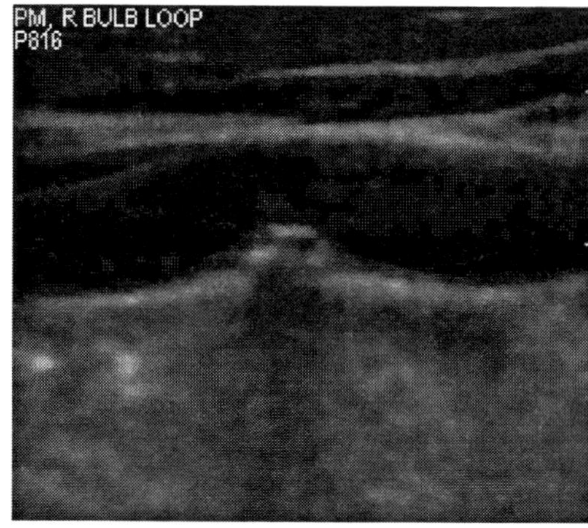

(b) Despeckled Frame 1 (C=0. 844)

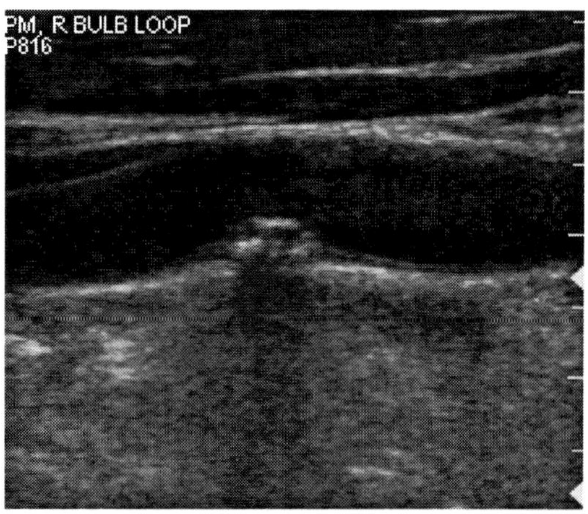

(c) Original Frame 100 (C=0.861)

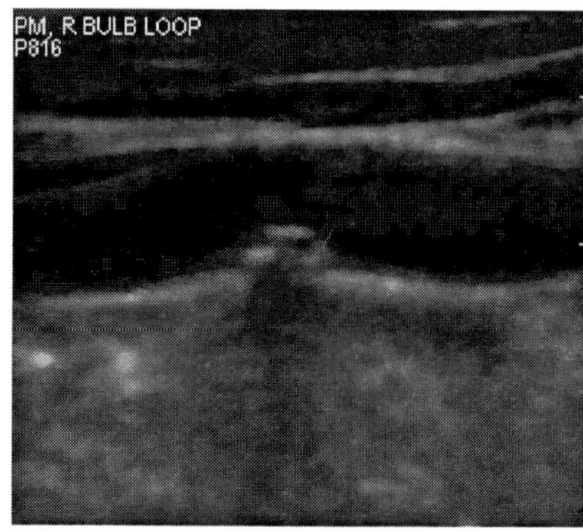

(d) Despeckled Frame 100 (C= 0.831)

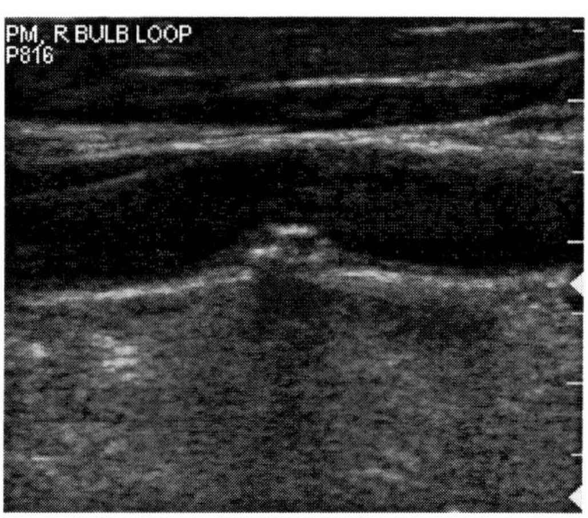

(e) Original Frame 200 (C=0.0855)

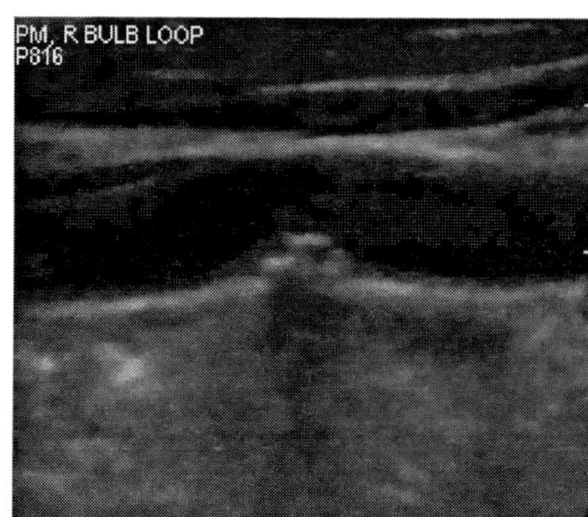

(f) Despeckled Frame 200 (C= 0.825)

Table 4.9 presents a comparison of the four different plaque snakes segmentation methods (Williams and Shah, Balloon, Lai and Chin, and GVF), on 80 longitudinal ultrasound images of the carotid plaque, based on the sensitivity, *R*, specificity, Sp, precision, *P*, and the measure *F*, as described in [60]. Bold values in Table 4.9 show best performance of the segmentation algorithms. The best sensitivity, *R*, was given by the Lai and Chin (0.827), followed by the Williams and Shah (0.8176), whereas the best specificity, Sp, was given by the Balloon (0.9460), followed by the Lai and Chin (0.9416) snakes segmentation method. The Lai and Chin gave the best precision, *P*, (0.9338), which is better than the rest of the segmentation methods, whereas the best *F*, was given by the Balloon (0.8882), followed by the Lai and Chin (0.8851) snakes segmentation method.

4.4.2 Video Despeckling

Figure 4.11 shows the original (see Figure 4.11a, c, and e) and despeckled (see Figure 4.11b, d, and f) frames 1, 150, and 300 from an ultrasound carotid artery video consisted out of 300 frames, with a width of 246 and a height of 256 pixels, and a total duration of 10 s (30 frames/s). The *DsFlsmv* despeckle filter was iteratively applied for three iterations at consecutive video frames with a moving sliding window [5 × 5] pixels. The filtering was only applied to the luminance channel of the video (Y-channel). The speckle index (C = ^{2}/) (3.13), for the original and despeckled frames, was also calculated, and it is given in parentheses. It is clear that *C* is reduced after despeckle filtering for all frames.

4.5 EVALUATION OF TWO DIFFERENT ULTRASOUND SCANNERS BASED ON DESPECKLE FILTERING

In this section, the results for the evaluation of image quality based on despeckle filtering performed on two different ultrasound imaging scanners (ATL HDI-3000 and ATL HDI-5000) are presented.

FIGURE 4.11: Despeckle filtering of carotid artery video for selected frames. The despeckle filter *DsFlsmv* was iteratively applied three times at each video frame, using a sliding moving window of size [5 × 5]. The carotid plaque is indicated with a box in the first frame at the far wall of the artery. (a) Original frame 1 (C = 0. 863). (b) Despeckled frame 1 (C = 0. 844). (c) Original frame 100 (C = 0.861). (d) Despeckled frame 100 (C = 0.831). (e) Original frame 200 (C = 0.0855). (f) Despeckled frame 200 (C = 0.825).

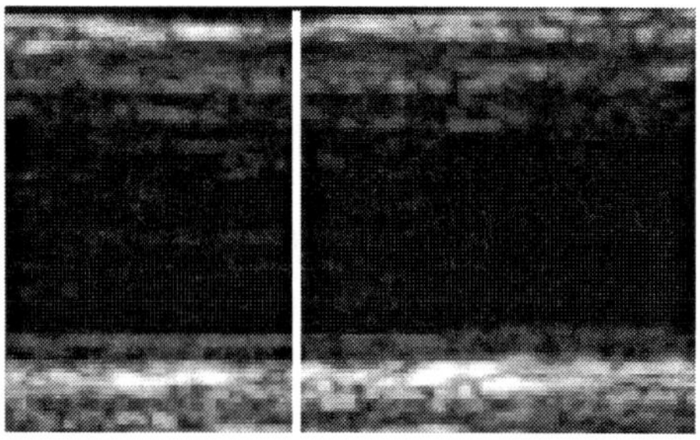

(a) Original (NF) 3000

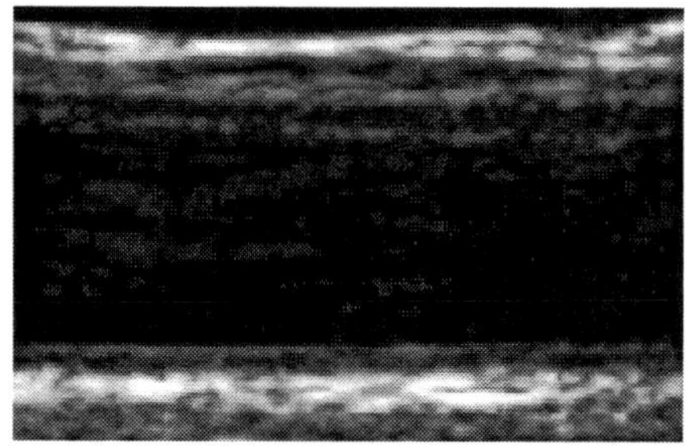

(b) Original (NF) 5000

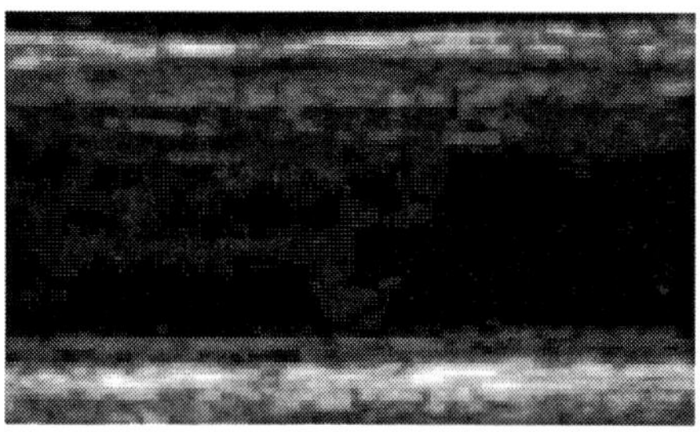

(c) Speckle reduction (DS) 3000

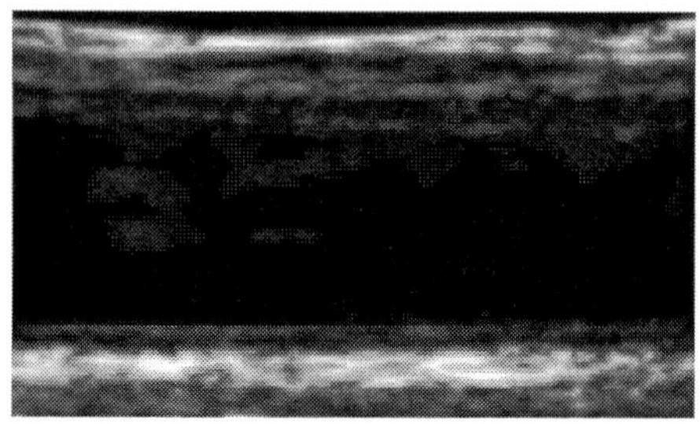

(d) Speckle reduction (DS) 5000

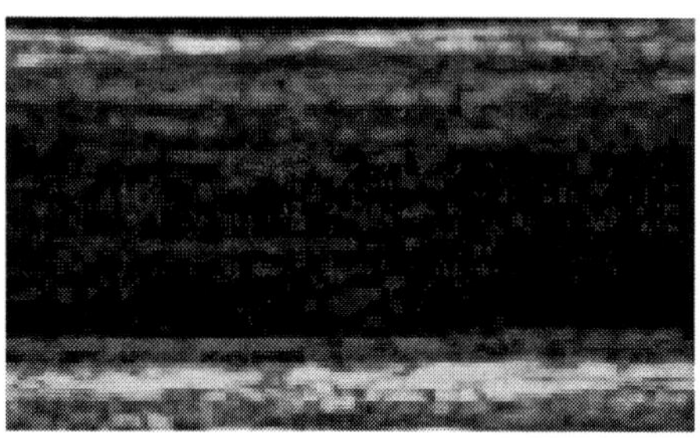

(e) Normalized (N) 3000

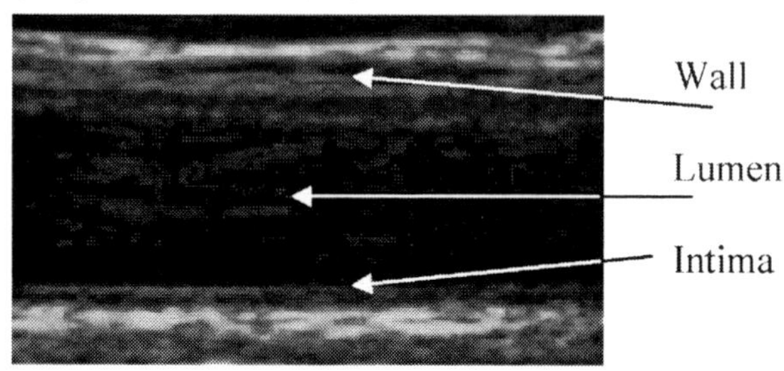

(f) Normalized (N) 5000

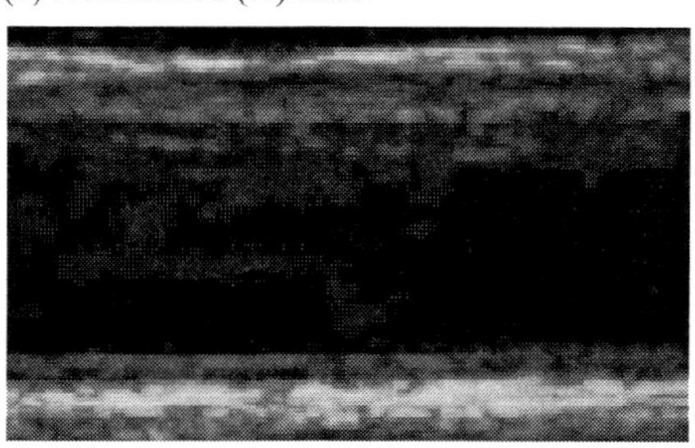

(g) Normalized speckle reduction (NDS) 3000

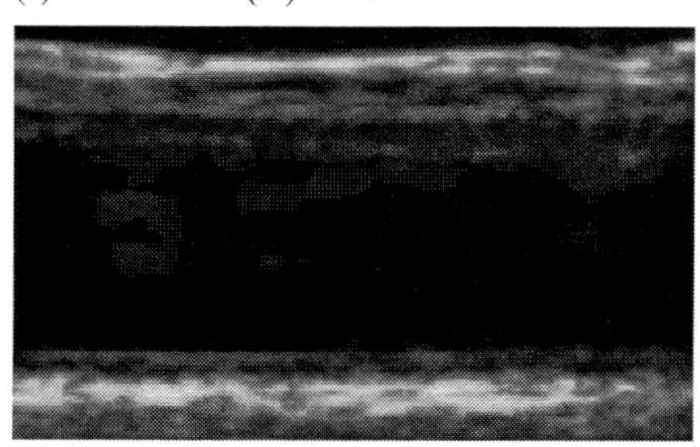

(h) Normalized speckle reduction (NDS)

4.5.1 Evaluation of Despeckle Filtering on an Ultrasound Image

Figure 4.12 illustrates the original, NF, despeckled, DS, normalized, N, and normalized despeckled, NDS, images for the two ultrasound image scanners. It is shown that the images for the ATL HDI-3000 scanner have greater speckle noise compared to the ATL HDI-5000 images. Moreover, the lumen borders and the IMT are more easily identified with the ATL HDI-5000 on the N and NDS images.

4.5.2 Evaluation of Despeckle Filtering on Gray-value Line Profiles

Figure 4.13 shows gray-value line profiles, from top to bottom of an ultrasound carotid image (see Figure 4.12a) for the original, NF, despeckled, DS, normalized, N, and normalized despeckled, NDS, images for the ATL HDI-3000 and ATL HDI-5000 scanner. Figure 4.13 also shows that speckle reduction filtering sharpens the edges. The contrast in the ATL HDI-3000 images was decreased after normalization and speckle reduction filtering, whereas the contrast for the ATL HDI-5000 images was increased after normalization.

4.5.3 Evaluation of Despeckle Filtering Based on Visual Perception Evaluation

Table 4.10 shows the results in percentage (%) format for the visual perception evaluation made by the two vascular experts on the two scanners. It is clearly shown that the highest scores are given for the NDS images, followed by the N, DS, and NF images for both scanners from both experts.

Table 4.11 presents the results of the Wilcoxon rank sum test for the visual perception evaluation, performed among the NF–DS, NF–N, NF–NDS, DS–N, DS–NDS, and N–NDS images, for the first and second observer on the ATL HDI-3000 and the ATL HDI-5000 scanner, respectively. The results of the Wilcoxon rank sum test in Table 4.11 for the visual perception evaluation were mostly significantly different (S) showing large intra-observer and inter-observer variability for the different preprocessing procedures (NF–DS, NF–N, NF–NDS, DS–N, DS–NDS, N–NDS)

FIGURE 4.12: Ultrasound carotid artery images of the original (NF), despeckle (DS), normalized (N), and normalized despeckled (NDS), of the ATL HDI-3000 and ATL HDI-5000 shown in the left and right columns, respectively. Vertical lines given in the original image (NF) of the ATL HDI-3000 and the ATL HDI-5000 scanners, define the position of the gray-value line profiles plotted in Figure 4.13. Source [59], © MBEC 2006. (a) Original (NF) 3000. (b) Original (NF) 5000. (c) Speckle reduction (DS) 3000. (d) Speckle reduction (DS) 5000. (e) Normalized (N) 3000. (f) Normalized (N) 5000. (g) Normalized speckle reduction (NDS) 3000. (h) Normalized speckle reduction (NDS).

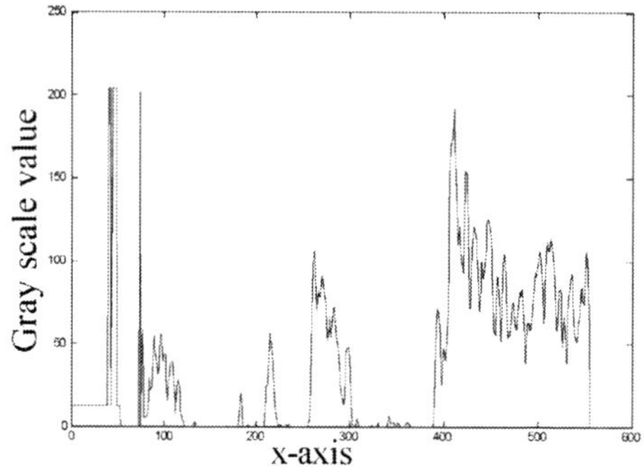

(a) Original (NF) 3000

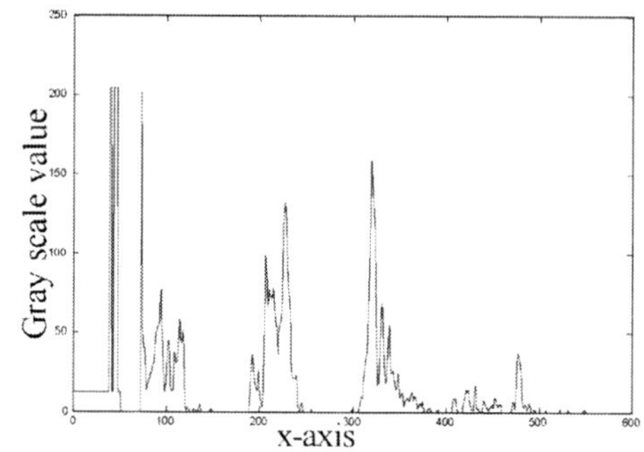

(b) Original (NF) 5000

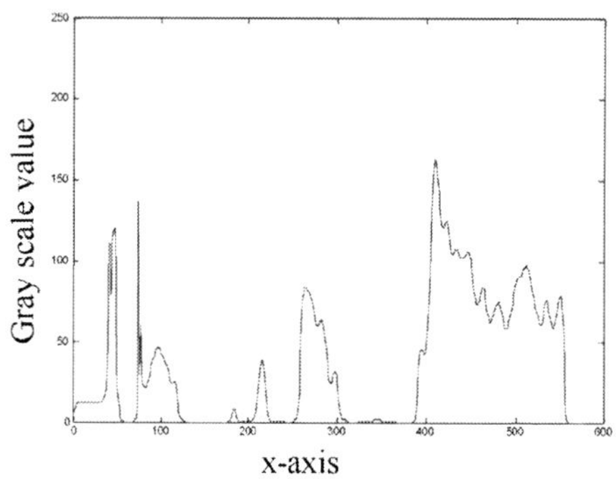

(c) Speckle reduction (DS) 3000

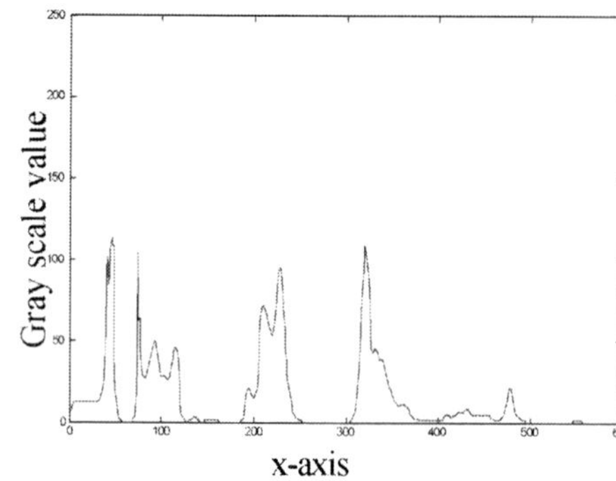

(d) Speckle reduction (DS) 5000

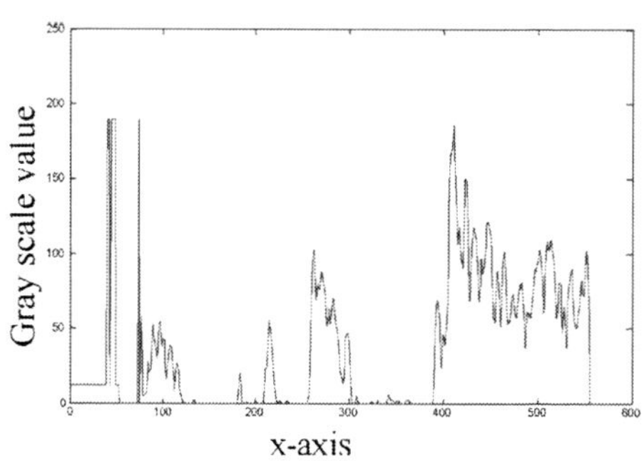

(e) Normalized (N) 3000

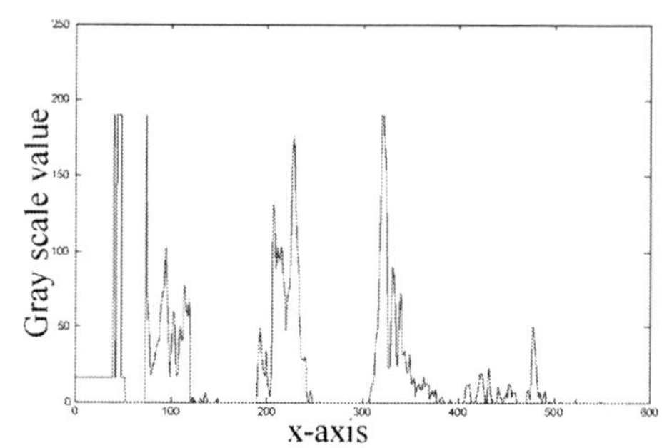

(f) Normalized (N) 5000

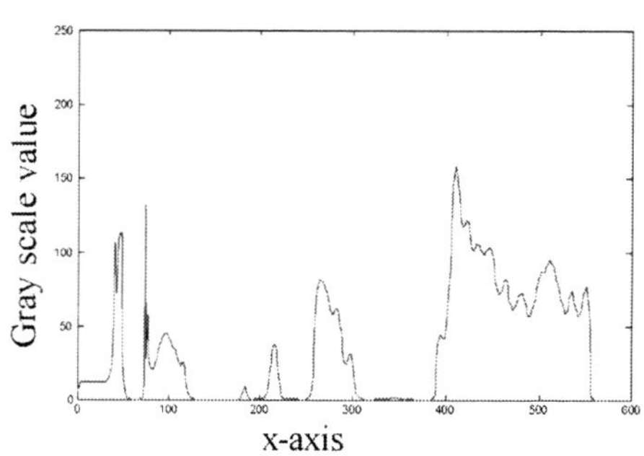

(g) Normalized speckle reduction (NDS) 3000

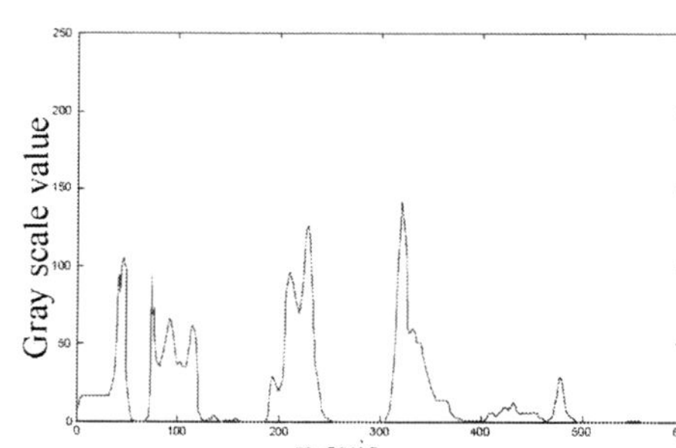

(h) Normalized speckle reduction (NDS) 5000

TABLE 4.10: Visual perception evaluation for the image quality on 80 images processed from each scanner for the original (NF), despeckled (DS), normalized (N), and normalized despeckled (NDS)

VISUAL PERCEPTION SCORE								
ULTRASOUND SCANNER	ATL HDI-3000				ATL HDI-5000			
Preprocessing procedure	NF	DS	N	NDS	NF	DS	N	NDS
Angiologist	30	43	69	72	26	42	59	70
Neurovascular specialist	41	56	54	71	49	53	59	72
Average	36	50	62	72	38	48	59	71

Source [59], © MBEC 2006.

for both scanners. Not significantly (NS) different values were obtained for both scanners, after normalization and speckle reduction filtering, showing that this improves the optical perception evaluation.

4.5.4 Evaluation of Despeckle Filtering based on Statistical and Texture Features

Table 4.12 presents the results of the statistical and texture features for the 80 images recorded from each image scanner. The upper part of Table 4.12 shows that the effect of speckle reduction filtering, DS, for both scanners was similar, that is, the mean and the median were preserved, the standard deviation was reduced, the skewness and the kurtosis were reduced, and the speckle index was reduced (see also Figure 4.10c, d, g and h, where it is shown that the gray-value line profiles are smoother and less flattened). Furthermore, Table 4.12 shows that some statistical measures like the skewness, kurtosis, and speckle index, were better than the original, NF, and speckle reduction, DS, images after normalization, N, for both scanners, and were even better after normalization

FIGURE 4.13: Gray-value line profiles of the lines illustrated in Figure 4.12a and b, for the NF, DS, N, and NDS images, for the ATL HDI-3000 and ATL HDI-5000 scanner, shown in the left and right columns, respectively. The gray-scale value and the column 240, is shown in the *y*- and *x*-axis. (a) Original (NF) 3000. (b) Original (NF) 5000. (c) Speckle reduction (DS) 3000. (d) Speckle reduction (DS) 5000. (e) Normalized (N) 3000. (f) Normalized (N) 5000. (g) Normalized speckle reduction (NDS) 3000. (h) Normalized speckle reduction (NDS) 5000.

TABLE 4.11: Wilcoxon rank sum test p value for the ATL HDI-3000 and the ATL HDI-5000 scanner for the visual perception evaluation performed by the experts among the NF–DS, NF–N, NF–NDS, DS–N, DS–NDS, and N–NDS images

ULTRASOUND SCANNER	ATL HDI-3000					
PREPROCESSING PROCEDURE	**NF–DS**	**NF–N**	**NF–NDS**	**DS–N**	**DS-NDS**	**N-NDS**
Angiologist	$1.2*10^{-4}$	$1.1*10^{-11}$	$1.1*10^{-11}$	$1.3*10^{-8}$	$1.1*10^{-8}$	0.385
	(S)	(S)	(S)	(S)	(S)	(NS)
Neurovascular specialist	$2.9*10^{-4}$	0.004	$3.5*10^{-9}$	0.55	$1.7*10^{-4}$	$1.5*10^{-4}$
	(S)	(S)	(S)	(NS)	(S)	(S)
Ultrasound scanner	ATL HDI-5000					
Preprocessing procedure	NF–DS	NF–N	NF–NDS	DS–N	DS–NDS	N–NDS
Angiologist	0.14	0.001	$9.6*10^{-8}$	0.65	$8.9*10^{-6}$	$7.6*10^{-8}$
	(NS)	(S)	(S)	(NS)	(S)	(S)
Neurovascular specialist	0.85	$1.3*10^{-4}$	$6.1*10^{-8}$	0.56	0.002	0.001
	(NS)	(S)	(S)	(NS)	(S)	(S)

The test shows in parenthesis with S significant difference at $p < 0.05$ and NS no significant difference at p 0.05. Source [59], © MBEC 2006.

and speckle reduction, NDS. However, the mean was increased for N and NDS images for both scanners.

In the bottom part of Table 4.12, it is shown that the entropy was increased, and the contrast was reduced significantly in the cases of DS and NDS for both scanners. The entropy was slightly increased, and the contrast was slightly reduced in the cases of N images for both scanners. The ASM was reduced for the DS images for both scanners and for the NDS images for the ATL HDI-5000 scanner.

Table 4.13 presents the results of the Wilcoxon rank sum test for the statistical and texture features (see Table 4.12), performed on the NF–DS, NF–N, NF–NDS, DS–N, DS–NDS, and N–NDS images on the ATL HDI-3000 scanner. No statistically significant difference was found in the first part of Table 4.13 when performing the nonparametric Wilcoxon rank sum test at $p < 0.05$, between

TABLE 4.12: Statistical and texture features (Mean values for 80 images processed from each scanner) for the original (NF), despeckled (DS), normalized (N) and normalized despeckled (NDS) images

SCANNER	ATL HDI-3000				ATL HDI-5000			
IMAGES	NF	DS	N	NDS	NF	DS	N	NDS
Statistical features (SF)								
Mean	22.13	21.78	26.81	26.46	22.72	22.35	27.81	27.46
Median	3.07	4.53	3.56	5.07	3.73	5.23	4.59	6.07
Stand. deviation	40.67	36.2	45.15	41.48	41.22	36.7	45.9	42.31
Skewness (^{3})	2.88	2.49	2.23	2.00	2.84	2.45	2.17	1.94
Kurtosis (^{4})	12.43	10.05	7.94	6.73	12.13	9.82	7.56	6.43
Speckle Index	0.29	0.27	0.25	0.24	0.28	0.27	0.24	0.23
SGLDM—range of values								
Entropy	0.24	0.34	0.25	0.34	0.40	0.48	0.41	0.48
Contrast	667	309	664	303	618	302	595	287
ASM	0.36	0.35	0.38	0.37	0.37	0.33	0.39	0.35

Source [59], © MBEC 2006.

the original, NF and despeckled, DS, the original, NF and normalized, N, and the original, NF and normalized despeckled, NDS features for both scanners. Statistical significant different values were mostly obtained for the second part of Table 4.13 for the ASM, contrast, and entropy.

Furthermore, Table 4.13 shows that the entropy that is a measure of the information content of the image was higher for the ATL HDI-5000 in all the cases. The ASM that is a measure of the inhomogeneity of the image is lower for the ATL HDI-5000 in the cases of the DS and NDS images. Furthermore, the entropy and the ASM were more influenced from speckle reduction than normalization, as they are reaching their best values after speckle reduction filtering.

4.5.5 Evaluation of Despeckle Filtering Based on Image Quality Evaluation Metrics

Table 4.14 illustrates the image quality evaluation metrics, for the 80 ultrasound images recorded from each image scanner, among the NF–DS, NF–N, NF–NDS, and N–NDS images. Best values were obtained for the NF–N images with lower RMSE, Err3, and Err4, higher SNR, and PSNR

TABLE 4.13: Wilcoxon rank sum test p value for the ATL HDI-3000 scanner for the statistical and texture features among the NF–DS, NF–N, NF–NDS, DS–N, DS–NDS, and N–NDS images

PREPROCESSING PROCEDURE	ATL HDI-3000					
	NF–DS	NF–N	NF–NDS	DS–N	DS–NDS	N–NDS
Statistical features (SF)						
Mean	0.69	0.5	0.07	0.56	0.31	0.09
	(NS)	(NS)	(NS)	(NS)	(NS)	(NS)
Median	0.02	0.09	0.07	0.001	0.34	0.03
	(S)	(NS)	(NS)	(S)	(NS)	(S)
Stand. deviation (^{2})	0.01	0.02	0.08	0.03	0.004	$3.8*10^{-4}$
	(S)	(S)	(NS)	(S)	(S)	(S)
Skewness (^{3})	0.08	0.45	$7.3*10^{-4}$	0.037	0.17	0.07
	(NS)	(NS)	(S)	(S)	(NS)	(NS)
Kurtosis (^{4})	0.08	0.09	$4.5*10^{-4}$	0.19	0.34	0.07
	(NS)	(NS)	(S)	(NS)	(NS)	(S)
SGLDM—range of values						
Entropy	$6.9*10^{-7}$	0.09	$2.2*10^{-3}$	$7.1*10^{-11}$	0.17	$4.2*10^{-5}$
	(S)	(NS)	(S)	(S)	(NS)	(S)
Contrast	$3*10^{-12}$	0.25	$4.2*10^{-7}$	$3.1*10^{-5}$	0.45	$5.6*10^{-9}$
	(S)	(NS)	(S)	(S)	(NS)	(S)
ASM	$9.6*10^{-7}$	$2.2*10^{-9}$	$1.4*10^{-6}$	$6.7*10^{-8}$	$7.2*10^{-7}$	$4.3*10^{-7}$
	(S)	(S)	(S)	(S)	(S)	(S)

The test shows in parenthesis with S significant difference at $p < 0.05$ and NS no significant difference at p 0.05. Source [59], © MBEC 2006.

for both scanners. The GAE was 0.00 for all cases, and this can be attributed to the fact that the information between the original and the processed images remains unchanged. Best values for Q and SSIN were obtained for the NF–N images for both scanners, whereas best values for SNR were obtained for the ATL HDI-3000 scanner on the NF–N images.

TABLE 4.14: Image quality evaluation metrics among the original-despeckled (NF–DS), original-normalized (NF–N), original-normalized despeckled (NF–NDS), and the normalized-normalized despeckled (N–NDS) images

EVALUATION METRICS	ATL HDI-3000				ATL HDI-5000			
	NF–DS	NF–N	NF–NDS	NF–NDS	NF–DS	NF–N	NF–NDS	N–NDS
MSE	1.4	1.3	2.0	1.3	1.2	0.3	1.9	1.3
RMSE	1.2	0.4	1.4	1.1	1.1	0.5	1.3	1.1
Err 3	3.8	0.8	3.9	3.5	3.7	0.8	3.8	3.5
Err 4	8.2	1.2	8.0	7.5	8.1	1.3	7.8	7.5
GAE	0	0	0	0	0	0	0	0
SNR	5.0	16.5	4.8	5.4	5.3	15.9	5.1	5.4
PSNR	48.0	59	45.6	44.6	47.4	58.5	46	44.6
Q	0.7	0.93	0.73	0.69	0.72	0.93	0.72	0.71
SSIN	0.9	0.95	0.92	0.83	0.94	0.95	0.91	0.83

Source [59], © MBEC 2006.

Table 4.14 shows that the effect of speckle reduction filtering was more obvious on the ATL HDI-3000 scanner, which shows that the ATL HDI-5000 scanner produces images with lower noise and distortion. Moreover, it was obvious that all quality metrics presented here are equally important for image quality evaluation. Specifically, for the most of the quality metrics, better measures were obtained between the NF–N, followed by the NF–NDS, and N–NDS images for both scanners. It is furthermore important to note that a higher PSNR (or equivalently, a lower RMSE) does not necessarily imply a higher subjective image quality, although they do provide some measure of relative quality.

Furthermore, the two experts evaluated visually 10 B-mode ultrasound images with different types of plaque [53] (see Figure 4.14), by delineating the plaque at the far wall of the carotid artery wall. The visual perception evaluation, and the delineations made by the two experts, showed that the plaque may be better identified on the ATL HDI-5000 scanner after normalization and speckle reduction, NDS, whereas the borders of the plaque and the surrounding tissue may be better visualized on the ATL HDI-5000 when compared with the ATL HDI-3000 scanner.

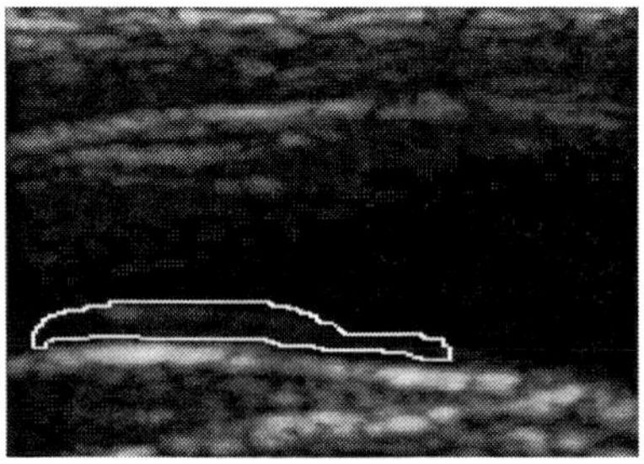

(a) Original (NF) 3000

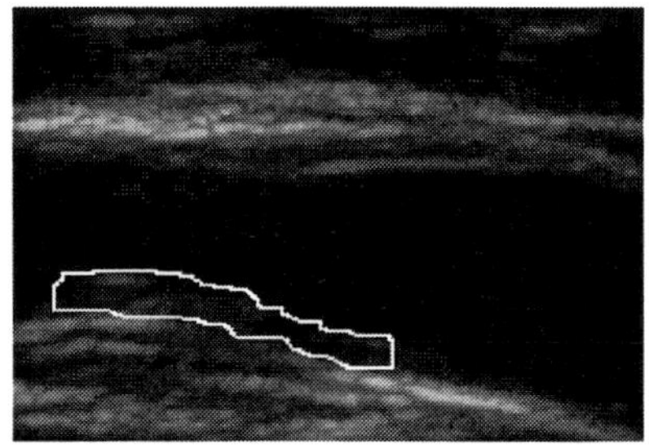

(b) Original (NF) 5000

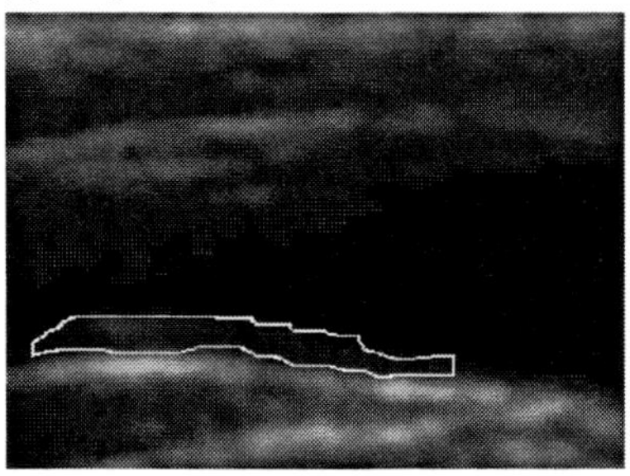

(c) Despeckled (DS) 3000

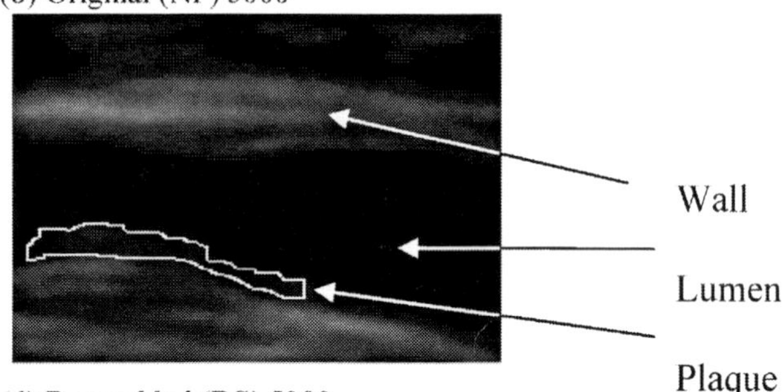

(d) Despeckled (DS) 5000

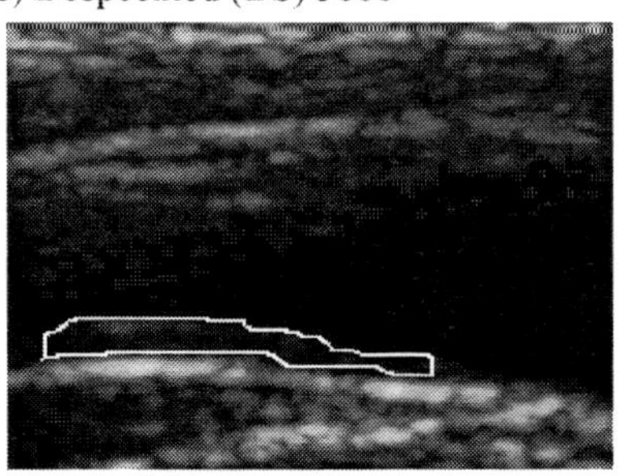

(e) Normalized (N) 3000

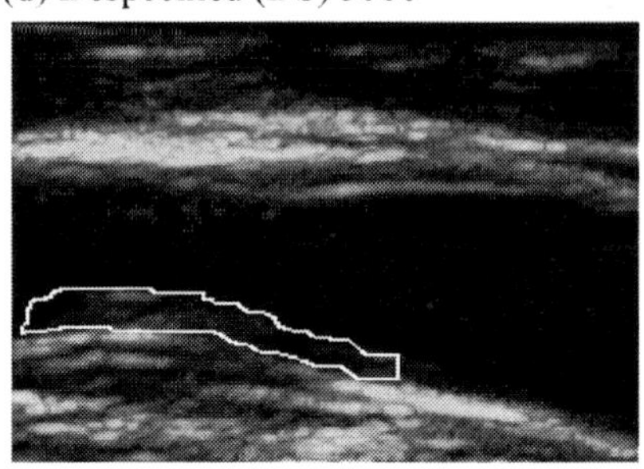

(f) Normalized (N) 5000

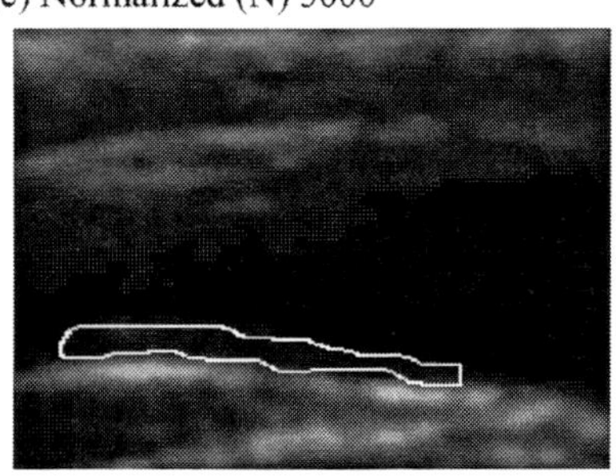

(g) Normalized despeckled (NDS) 3000

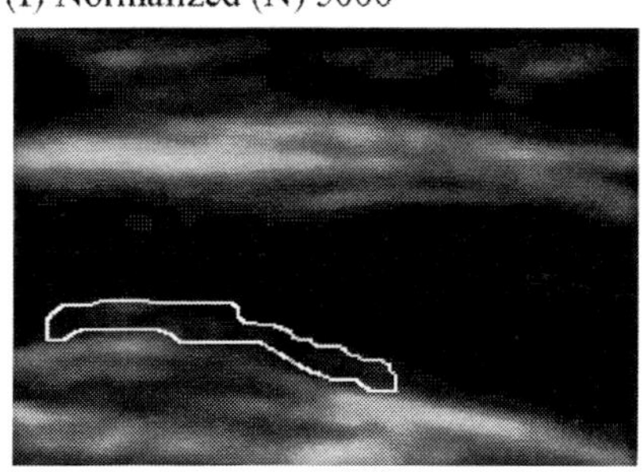

(h) Normalized despeckled (NDS) 5000

TABLE 4.15: Summary findings of image quality evaluation in ultrasound imaging of the carotid artery

ULTRASOUND SCANNER	VISUAL EVALUATION TABLE 4.10				STATISTICAL AND TEXTURE ANALYSIS TABLES 4.12–4.13				IMAGE QUALITY EVALUATION TABLE 4.14			
	NF	DS	N	NDS	NF	DS	N	NDS	NFDS	NF-N	NF-NDS	N-NDS
ATL HDI-3000				++		+		++		++	+	
ATL HDI-5000				++		+		++	+	++		

Table 4.15 summarizes the image quality evaluation results of this study, for the visual evaluation (Table 4.10), the statistical and texture analysis (Table 4.12), and the image quality evaluation metrics (Table 4.14). A double plus sign in Table 4.15 indicates very good performance, while a single plus sign a good performance. Table 4.15 can be summarized as follows: (i) the NDS images were rated visually better on both scanners, (ii) the NDS images showed better statistical and texture analysis results on both scanners, (iii) the NF–N images on both scanners showed better image quality evaluation results, followed by the NF-DS on the ATL HDI-5000 scanner and the NF–DS on the HDI ATL-3000 scanner, (iv) the ATL HDI-5000 scanner images have considerable higher entropy than the ATL HDI-3000 and thus more information content. However, based on the visual evaluation by the two experts, both scanners were rated similarly.

* * * *

FIGURE 4.14: Ultrasound carotid plaque images of Type II outlined by an expert of the original (NF), speckle reduction (DS), normalized (N), and normalized speckle reduction (NDS), of the ATL HDI-3000 and ATL HDI-5000 shown in the left and right columns, respectively. Source [59], © MBEC 2006. (a) Original (NF) 3000. (b) Original (NF) 5000. (c) Despeckled (DS) 3000. (d) Despeckled (DS) 5000. (e) Normalized (N) 3000. (f) Normalized (N) 5000. (g) Normalized despeckled (NDS) 3000. (h) Normalized despeckled (NDS) 5000.

CHAPTER 5

Comparison and Discussion of Despeckle Filtering Algorithms

Despeckle filtering is an important operation in the enhancement of ultrasound images of the carotid artery, both in the case of texture analysis, and in the case of image quality evaluation and visual evaluation by the experts. In this study, a total of 10 despeckle filters were comparatively evaluated on 440 ultrasound images of the carotid artery bifurcation, and the validation results are summarized in Table 5.1.

As given in Table 5.1, filters *DsFlsmv*, *DsFlsminsc*, and *DsFhomo*, improved the class separation between the asymptomatic and the symptomatic classes (see also Table 4.2). Filters *DsFlsmv*, *DsFlsminsc*, and *DsFgf4d* gave a high number of significantly different features (see Table 4.3). Filters *DsFlsminsc*, *DsFgf4d*, and *DsFhomo* gave only a marginal improvement in the percentage of correct classifications success rate (see Table 4.4). Moreover, filters *DsFlsmv*, *DsFnldif*, and *DsFwaveltc* gave better image quality evaluation results (see Table 4.5). Filters *DsFlsmv* and *DsFgf4d* improved the visual assessment carried out by the experts (see Table 4.6). It is clearly shown that filter *DsFlsmv* gave the best performance, followed by filters *DsFlsminsc* and *DsFgf4d* (see Table 5.1). Filter *DsFlsmv* or *DsFgf4d* can be used for despeckling asymptomatic images where the expert is interested mainly in the plaque composition and texture analysis. Filters *DsFlsmv* or *DsFgf4d* or *DsFlsminsc* can be used for despeckling of symptomatic images where the expert is interested in identifying the degree of stenosis and the plaque borders. Filters *DsFhomo*, *DsFnldif*, and *DsFwaveltc* gave poorer performance.

Filter *DsFlsmv* gave very good performance, with respect to: (i) preserving the mean and the median as well as decreasing the variance and the speckle index of the image, (ii) increasing the distance of the texture features between the asymptomatic and the symptomatic classes, (iii) significantly changing the SGLDM range of values texture features after filtering based on the Wilcoxon rank sum test, (iv) marginally improving the classification success rate of the kNN classifier for the classification of asymptomatic and symptomatic images in the cases of SF, SMF, and TEM feature sets, and (v) improving the image quality of the image. The *DsFlsmv* filter, which is a simple filter, is based on local image statistics. It was first introduced in Refs. [14], [9], [15] by Jong-Sen Lee and

TABLE 5.1: Summary findings of despeckle filtering in ultrasound imaging of the carotid artery

DESPECKLE FILTER	STATISTICAL AND TEXTURE FEATURES TABLE 4.2	STATISTICAL ANALYSIS TABLE 4.3	KNN CLASSIFIER TABLE 4.4	IMAGE QUALITY EVALUATION TABLE 4.5	OPTICAL PERCEPTION EVALUATION TABLES 4.6A, B
Linear filtering					
DsFlsmv	✓	✓		✓	✓
DsFlsminsc	✓	✓	✓		
Nonlinear filtering					
DsFgf4d		✓	✓		✓
DsFhomo	✓		✓		
Diffusion filtering					
DsFnldif				✓	
Wavelet filtering					
DsFwaveltc				✓	

coworkers, and it was tested on a few SAR images with satisfactory results. It was also used for SAR imaging in Ref. [13] and image restoration in Ref. [16], again with satisfactory results.

Filter *DsFlsminsc* gave the best performance with respect to: (i) preserving the mean, as well as decreasing the variance and the speckle index and increasing the contrast of the image, (ii) increasing the distance of the texture features between the asymptomatic and the symptomatic classes, (iii) significantly changing the SGLDM texture features after filtering based on the Wilcoxon rank sum test, (iv) improving the classification success rate of the kNN classifier for the classification of asymptomatic and symptomatic images in the cases of SF, SGLDMr, GLDS, NGTDM, FDTA, and FPS feature sets. Filter *DsFlsminsc* was originally introduced by Nagao in Ref. [32] and was tested on an artificial and a SAR image with satisfactory performance. In this study, the filter was modified, by using the speckle index instead of the variance value for each subwindow [as described in Section 2.1.3, Eq. (2.10), see also Eq. (3.13)].

Filter *DsFgf4d* gave very good performance with respect to: (i) decreasing the speckle index, (ii) marginally increasing the distance of the texture features between the asymptomatic and the symptomatic classes, (iii) significantly changing the SGLDM range of values texture features after filtering based on the Wilcoxon rank sum test, (iv) improving the classification success rate of the kNN classifier for the classification of asymptomatic and symptomatic images in the cases of SGLDMm, GLDS, NGTDM, SFM, and TEM feature sets. The geometric filter *DsFgf4d* was introduced by Crimmins [10] and was tested visually on a few SAR images with satisfactory results.

5.1 COMPARISON AND DISCUSSION OF DESPECKLE FILTERING ALGORITHMS

Phantom images were used in this book (see Figure 4.1) and by other researchers to evaluate despeckle filtering in carotid ultrasound imaging. Specifically in Ref. [24], a synthetic carotid ultrasound image of the CCA was used to evaluate the *DsFsard* filtering (speckle-reducing anisotropic filtering) which was compared with the *DsFlsmv* (Lee filter) [14] and the *DsFad* filter (conventional anisotropic diffusion) [22]. The edges of the phantom image used in Ref. [24] were studied, and it was shown that the *DsFsrad* does not blur edges as with the other two despeckle filtering techniques evaluated (*DsFlsmv* and *DsFad*).

Despeckle filtering was investigated by other researchers and also in our study, on an artificial carotid image (Figure 4.4) [5, 7] on line profiles (Figure 4.5) of different ultrasound images [7, 8, 19, 24, 28], on phantom ultrasound images (Figures 4.1–4.3) [24 26, 31], SAR images [17, 34, 77, 78], real longitudinal ultrasound images of the carotid artery (Fig. 4.6) [5, 7, 24] and cardiac ultrasound images (see Figure 4.7). There are only two studies, Refs. [5] and [7], where despeckle filtering was investigated on real and artificial longitudinal ultrasound image of the carotid artery. Four different despeckle filters were applied in Ref. [5], namely, the *DsFlsmv* [14], Frost [13], *DsFad* [23], and a *DsFsrad* filter [5]. The despeckle window used for the *DsFlsmv* and Frost filters was [7 × 7] pixels. To evaluate the performance of these filters, the mean and the standard deviation were used, which were calculated in different regions of the carotid artery image, namely, in lumen, tissue, and at the vascular wall. The mean gray-level values of the original image for the lumen, tissue, and wall regions were 1.03, 5.31, and 22.8, whereas the variances were 0.56, 2.69, and 10.61. The mean after despeckle filtering with the *DsFsrad* gave brighter values for the lumen and tissue. Specifically, the means for the lumen, tissue, and wall for the *DsFsrad* were (1.19, 6.17, 18.9), the *DsFlsmv* were (1.11, 5.72, 21.75), the Frost were (1.12, 5.74, 21.83), and the *DsFad* were (0.90, 4.64, 14.64). The standard deviation for the *DsFsrad* gave lower values (0.15, 0.7, 2.86) when compared with Lee (0.33, 1.42, 5.37), Frost (0.32, 1.40, 5.30), and *DsFad* (0.20, 1.09, 3.52). It was thus shown that the *DsFsrad* filter preserves the mean and reduces the variance. The number of images investigated in Ref. [5] was very small, visual perception evaluation by experts was not carried out, and only two statistical measures were used to quantitatively evaluate despeckle filtering, namely, the mean and the variance before and after despeckle filtering as explained above. We believe that the mean and the variance used in Ref. [5] are not indicative and may not give a complete and accurate evaluation result as in Ref. [7]. Furthermore, despeckle filtering was investigated by other researchers on ultrasound images of heart [24] (see also Figure 4.7), pig heart [28], pig muscle [62], kidney [26], liver [63], echocardiograms [27], CT lung scans [21], MRI images of brain [64], brain X-ray images [65], SAR images [17], and real-world images [66].

Line plots, as used in our study (see Figure 4.5), were also used in a few other studies to quantify despeckle filtering performance. Specifically in Ref. [65], a line profile through the original and the despeckled ultrasound image of kidney was plotted, using adaptive Gaussian filtering. In Ref. [67], line profiles were plotted on four simulated and 15 ultrasound cardiac images of the left ventricle, to evaluate the *DsFmedian* filter. In another study [24], line profiles through one phantom, one heart, one kidney, and one liver ultrasound image, were plotted where an adaptive shrinkage weighted median [63, 66], *DsFwaveltc* (wavelet shrinkage) [29], and wavelet shrinkage coherence-enhancing [27] models were used and compared with a nonlinear coherent diffusion model [29]. Finally in Ref. [28], line plots were used in one artificial computer-simulated image and one ultrasound image of pig heart, where an adaptive shrinkage weighted median filter [63, 66], a multiscale nonlinear thresholding without adaptive filter preprocessing [28], a wavelet shrinkage filtering method [29], and a proposed adaptive nonlinear thresholding with adaptive preprocessing method [28] were evaluated. In all of the above studies, visual perception evaluation by experts, statistical and texture analysis, on multiple images, as performed in our study, were not performed.

5.2 DESPECKLE FILTERING OF CAROTID PLAQUE IMAGES BASED ON TEXTURE ANALYSIS

The results on texture analysis, presented in Chapter 4 (Tables 4.2–4.4), showed that the filters, *DsFlsmv*, *DsFgf4d*, and *DsFlsminsc* (Table 5.1), improved the class separation between the asymptomatic and the symptomatic classes (Table 4.2), by increasing the distance between them. These filters, *DsFlsmv*, *DsFgf4d*, and *DsFlsminsc*, gave the highest number of significantly different features (Table 4.3), with 7, 6, and 5, respectively, and gave only a marginal improvement in the percentage of correct classification success rate (Table 4.4). The high number of significantly different features for these filters showed that the two classes (asymptomatic, symptomatic) may be better separated after despeckle filtering with the filters *DsFlsmv*, *DsFgf4d*, and *DsFlsminsc*. Table 4.3 showed that almost all despeckle filters increased the distance between the asymptomatic and the symptomatic images thus making the identification of a class easier to identify. Table 4.3 also showed that most of the filters reduced the asymmetry, ^{3}, and the skewness, ^{4}, of the histogram. Table 4.4 showed that despeckle filtering influenced more some statistical features, such as the inverse difference moment, IDM, the angular second moment, ASM, and the sum entropy, *Entr*, while other statistical features were less influenced by despeckle filtering. As a result, these features, which were more influenced, may be used in future research to evaluate despeckle filtering. The *Score_Dis_T* in the last row of Table 4.3 showed that best feature distance was given by the filters *DsFhomo*, *DsFlsminsc*, *DsFmedian*, and *DsFlsmv*. Table 4.4 showed that not all feature sets equally benefited from despeckle filtering. Specifically, the SF and TEM feature sets benefited from almost all despeckle filters (7), whereas the feature sets SGLDMm, GLDS, and NGTDM, benefited from four despeckle filters, three FDTA, and two SFM. The features sets, SGLDMr and FPS, benefited from only one despeckle filter.

There were some results given in the recent literature based on texture analysis of ultrasound images for the classification of atherosclerotic carotid plaque [7, 8, 40, 68], liver ultrasound images [44], electron microscopic muscle images [69], detection of breast masses [70], cloud images [71], SAR images [17, 34], and some results given on artificial images from the pioneer researchers in texture analysis [41, 42]. There is no other study reported in the literature, where texture analysis (Tables 4.2–4.4) was used to the extent, that is used in our study, to evaluate despeckle filtering in ultrasound imaging. In studies [71], [72], some of the texture measures used in our study (Table 4.3), were also used on a total of 230 ultrasound images of the carotid plaque (115 asymptomatic, 115 symptomatic), to characterize carotid plaques as safe or unsafe and identify patients at risk of stroke. Specifically in Refs. [44] and [71], all nine different features used in our study (see Table 4.4) were also used to classify a plaque as asymptomatic or symptomatic, where comparable values as in our study were obtained for all feature sets. Examples of the use of texture analysis were also provided in Ref. [73], for classifying malignant and benign tumours of breast, in Ref. [71], for classifying clouds and predicting weather, and finally in Ref. [42], to automatically classify terrain texture.

5.3 DESPECKLING OF THE INTIMA–MEDIA COMPLEX AND THE PLAQUE

In a recent study [54] performed on 100 ultrasound images of the CCA, normalization and speckle reduction filtering was used as a preprocessing step before segmentation of the IMT, based on our previous work [7], where it was shown that this improves the image quality, the visual evaluation of the image [59], and the outcome of the IMT segmentation [54]. Results of this study were also presented in Section 4.4 of this book. More specifically in studies [54] and [55], IMT segmentation was performed on ultrasound images of the CCA, and it was shown that normalization and despeckle filtering improves the outcome of the IMT segmentation algorithm and produces more accurate and reproducible results when compared with the manual segmentation method. Speckle reduction filtering of the carotid artery was also applied in Ref. [24], where it was also shown that this improves the image quality and visual interpretation of the experts. More, specifically, in Ref. [7], it was shown that image normalization followed by speckle-reduction filtering produces better quality images, whereas the reverse (speckle-reduction filtering followed by normalization) might produce distorted edges. The preferred method is to apply first normalization and then speckle-reduction filtering for better results. Speckle-reduction filtering of the carotid was also proposed by Refs. [7, 54, 59] and [60] where it was shown that this improves the image quality and the visual evaluation of the image. However, in other segmentation studies for extracting the carotid artery plaque borders in intravascular ultrasound (IVUS) imaging, speckle was used as useful information [102].

Therefore, the preprocessing of ultrasound images of the carotid artery with normalization and speckle reduction, followed by the snakes initialization and the Williams and Shah segmentation algorithm can be used successfully in the measurement of IMT complementing the manual measurements. It should be furthermore noted that both the manual and the snakes segmentation measurements were performed on linear segments of the carotid artery because, in the clinical praxis, the experts are delineating the IMT only in those parts of the vessel where there are no significant artefacts, signal drop outs, and structure irregularities. The validity of the measurements of the proposed methodology can always be easily assessed by the vascular expert.

The comparison of the four different plaque snakes segmentation methods, proposed in our recent study [60] and also presented in Section 4.4, for the segmentation of the atherosclerotic carotid plaque from ultrasound images, showed that the Lai and Chin snakes segmentation method gave slightly better results, although these results were not statistically significant when compared with the other three snakes segmentation methods (Williams and Shah, Balloon, and GVF).

To the best of our knowledge, no other study carried out ultrasound image normalization as described in this study, before segmentation of the atherosclerotic carotid plaque. However, in Ref. [103], histogram equalization was performed on carotid artery ultrasound images for increasing the image contrast. The normalization method proposed in this book was documented to be helpful in the manual contour extraction and in the snake's segmentation of the IMT [54, 59] and plaque [60]. Moreover, this method increased the classification accuracy of different plaque types as assessed by the experts [93, 94].

In Section 4.3, it was shown that speckle reduction filtering can be used as a preprocessing step based on our previous work [7], [59]. More, specifically, in Ref. [59], it was shown that image normalization followed by speckle reduction produces better quality images, whereas the reverse (speckle reduction followed by normalization) might produce distorted edges. Speckle-reduction filtering of the carotid was also proposed by Loizou et al. in 2005, Abd-Elmoniem, et al., Loizou et al. in 2006, and Loizou et al. in 2007 [7, 24, 59, 60], where it was shown that this improves the image quality and the visual evaluation of the image. However, in other segmentation studies for extracting the carotid artery plaque borders in IVUS imaging, speckle was used as useful information [98, 99].

5.4 VIDEO DESPECKLING

Most of the papers published in the literature for video filtering are limited to the reduction of additive noise, mainly by frame averaging. More specifically, in Ref. [121], the Wiener filtering method was applied to 3D image sequences for filtering additive noise, but results have not been thoroughly discussed and compared with other methods. The method was superior when compared to the purely temporal operations implemented earlier [122]. The pyramid thresholding method was used in Ref. [122], and wavelet-based additive denoising was used in Ref. [123] for additive noise reduction in image sequences. In another study [120], the image quality and evaluation metrics were used for evaluating the additive noise filtering and the transmission of image sequences through telemedicine channels. An improvement of almost all the quality metrics extracted from the original and processed images was demonstrated. An additive noise reduction algorithm, for image sequences, using variance characteristics of the noise was presented in Ref. [124]. Estimated noise power and sum of absolute difference employed in motion estimation were used to determine the temporal filter coefficients. A noise measurement scheme using the correlation between the noisy input and the noise-free image was applied for accurate estimation of the noise power. The experimental results showed that the proposed noise reduction method efficiently removes noise. An efficient method for movie denoising that does not require any motion estimation was presented in Ref. [125]. The method was based on the fact that averaging several realizations of a random variable reduces the variance. The method was unsupervised and was adapted to denoise image sequences with an additive white noise while preserving the visual details on the movie frames. Very little attention has been paid to the problem of missing data (impulsive distortion) removal in image sequences. In Ref. [126], a 3D median filter for removing impulsive noise from image sequences was developed. This filter was implemented without motion compensation, and so the results did not capture the full potential of these structures. Furthermore, the median operation, although quite successful in the additive noise filtering in images, invariably introduces distortion when filtering of image sequences [126]. This distortion primarily takes the form of blurring fine image details.

The basic principles of despeckle filtering for still images presented in Chapter 2, i.e., the proposed despeckle filtering algorithms and the extraction of texture features, image quality evaluation metrics, and the optical perception evaluation procedure, can also be applied to video. The application of despeckle filters (see Chapters 2.1–2.4), the extraction of texture features (see Section 3.5), the calculation of image quality metrics (see Section 3.9), and the visual perception evaluation by experts (see Section 3.10) may also be applied to video. The video can be broken into frames, which can then be processed one by one and then grouped together to form the processed video. Preliminary results for the application of despeckle filtering in ultrasound carotid and cardiac video were presented in Chapter 4. However, significant work still remains to be carried out.

5.4.1 Discussion

In a recent Ph.D. dissertation, different approaches for Gaussian and impulse noise filtering in video sequences were investigated, where a motion-detail adaptive filter, and a fuzzy logic motion-detection filter for video denoising were proposed for both gray and color image sequences [129]. The first one computes weighted averaging of pixel values in a local window, where the weights are computed based on locally estimated motion and spatial detail. It builds on the ideas underlying two existing techniques, *k*-nearest neighbor filtering and threshold averaging, but combines these ideas in a novel fashion. The fuzzy logic method is based on a recursive temporal filter where the degree of filtering is determined based on a motion confidence estimate produced by a fuzzy logic motion detector. The recursive temporal filter is followed by a fuzzy logic spatial filter, which deals well with the nonstationary noise at the temporal filter's output. The fuzzy logic framework is shown as more robust than the binary equivalent of the method.

The two video despeckling algorithms were evaluated on video sequences with added Gaussian and impulse noise based on the peak SNR (PSNR) between the original and the despeckled frame/s. In addition, they have been tested on real television sequences and sequences recorded by surveillance cameras, where usually additive Gaussian or correlated noise is present. The visual results on the processed sequences showed good denoising performance of the proposed algorithms in case of Gaussian (additive) noise, whereas for sequences containing correlated noise, the performance was lower but the quality of noisy sequences was still noticeably improved.

5.5 IMAGE QUALITY AND VISUAL EVALUATION

The image quality evaluation results presented in Table 4.5 showed that the best values were obtained by the despeckle filters *DsFnldif*, *DsFlsmv*, and *DsFwaveltc*. It was shown from Table 4.5 that the effect of despeckle filtering was more obvious on the asymptomatic images, where generally better image quality evaluation results were obtained. Moreover, it is obvious that all quality evaluation metrics presented here were equally important for image quality evaluation. It is furthermore important to note that a higher PSNR (or equivalently, a lower RMSE) does not necessarily imply a higher subjective image quality, although they do provide some measure of relative quality. While some quality metrics for different images have been studied and proposed in the literature, such as for MRI [74], natural and artificial images [47], to the best of our knowledge, no other comparative study exists except that of Loizou et al. [7], which have investigated the application of the above metrics together with visual perception evaluation, on ultrasound images of the carotid artery. In previous studies [9, 12, 14, 17, 19], researchers evaluated image quality on real-world images using either only the visual perception by experts or some of the evaluation metrics presented in Table 4.5. In all these studies, the comparison of the proposed method was made with another one, based on image quality evaluation metrics, such as the MSE [7, 24, 26, 28, 66], PSNR [66], SNR [7, 65], C [7, 26], the mean, and the variance [5, 7, 17, 62, 63] and line plots [7, 14, 24, 28, 62] between the original and despeckled images. The usefulness of these measures was not investigated for the despeckling of ultrasound images. Furthermore, normalization and despeckling was not taken into consideration as in our study. In a recent study [59], we have investigated the image quality on ultrasound images of the carotid artery, where it was shown that despeckle filtering increases the quality of these images and also increases the accuracy of the IMT [54] and plaque [60] segmentation.

Image quality metrics were also investigated for the evaluation of ultrasound spatial compound scanning [75], to compare the quality of JPEG images before and after compression using the PSNR and SSIN [47], where values for the PSNR and SSIN of 8.45, and 0.96, were measured, respectively, while in our study, we have achieved values of 39 and 0.97, with the *DsFlsmv* filter (see Table 4.5). In Ref. [45], real-world images were evaluated based on their compression ratios, by using the MSE and *Q*, where values of 30 and 0.92 were reported, respectively. Furthermore, real-world images were also evaluated in Ref. [45], based on the MSE and *Q*, before and after histogram equalization (1144.2, 0.74), median filtering (14.47, 0.78), wavelet compression (16.03, 0.68), and spatial displacement (141.2, 0.5).

In another study [66], where various median filtering techniques were investigated on real-world images, the image quality measures, MSE and PSNR, were used to compare between the original and the filtered images. In Ref. [76], a number of quality metrics were reviewed to evaluate JPEG compression on still real-world images, such as the MSE, SNR, PSNR, M3, and M4. In Ref. [24], where despeckle filtering was investigated on artificial and ultrasound images of heart, kidney and abdomen, the MSE values reported after despeckle filtering were 289, 271, 132, and 121, for four different despeckle filtering methods, namely, the adaptive weighted median filtering [66], wavelet shrinkage-enhanced [27], wavelet shrinkage [23], and nonlinear coherence diffusion method [29]. Most of the researchers used the image quality measures such as the MSE [7, 24, 26, 28, 66], SNR [3, 7, 8, 25, 30], and PSNR [66], to compare the original with the despeckled images.

In the research of A. Achim, et al. [26], values reported for the MSE were 133, 43, 49, 26, 22 for the original, and four despeckled SAR images, respectively. In Achim‘s research, four different despeckling methods were used, namely, the Lee [9], gamma MAP filter [77], soft thresholding, and the WIN-SAR filter [26], which used a [7×7] pixel filtering window and were applied on real-world and SAR images.

In another study [26], MSE values reported were 26 for the original kidney ultrasound image, 13.7 after despeckling by median filtering [66], 13.8 after homomorphic *DsFwiener* filtering [17], 13.6 after soft thresholding [29], 13.5 after hard thresholding [29], and 12.74 after Bayesian denoising [26]. In our study, the MSE values for the filter *DsFlsmv*, *DsFwiener*, *DsFnldif*, and

DsFwaveltc, (Table 4.5) were 13, 19, 8, 11, for the asymptomatic, and 33, 44, 8, 23, for the symptomatic images, respectively, which are better or comparable with other studies reported above.

Normalization and speckle reduction filtering are very important preprocessing steps in the assessment of atherosclerosis in ultrasound imaging. The usefulness of image quality evaluation, in 80 ultrasound images of the carotid bifurcation, based on image quality metrics and visual perception after normalization and speckle-reduction filtering using two different ultrasound scanners (ATL HDI-3000 and ATL HDI-5000) was addressed in Section 4.9 and, furthermore, was discussed here in this chapter. Specifically, the images were evaluated, before and after speckle reduction, after normalization, and after normalization and speckle-reduction filtering (see Figure 4.12). The evaluation was based on visual evaluation by two experts (see Table 4.10), statistical and texture features (see Tables 4.11–4.13), image normalization, speckle reduction, as well as based on image quality evaluation metrics (see Table 4.14). It is noted that to the best of our knowledge, there are no other studies found in the literature for evaluating ultrasound image quality, based on speckle-reduction filtering and normalization performed on carotid artery images, acquired by two different ultrasound scanners.

The main findings of this study can be summarized as follows: (1) the NDS images were rated visually better on both scanners, (2) the NDS images showed better statistical and texture analysis results on both scanners, (3) better image quality evaluation results were obtained between the NF–N images for both scanners, followed by the NF–DS images for the ATL HDI-5000 scanner and the NF–DS on the HDI ATL-3000 scanner, (4) the ATL HDI-5000 scanner images have considerable higher entropy than the ATL HDI-3000 scanner and thus more information content. However, based on the visual evaluation by the two experts, both scanners were rated similarly.

It was shown that normalization and speckle reduction produces better images. Normalization was also proposed in other studies using blood echogenecity as a reference and applied in carotid artery images [75]. In Refs. [53, 59, 60], it was shown that normalization improves the image comparability by reducing the variability introduced by different gain settings, different operators, and different equipment. It should be noted that the order of applying these processes (normalization and speckle reduction filtering) affects the final result. Based on unpublished results, we have observed that by applying first speckle-reduction filtering and then normalization produces distorted edges. The preferred method is to apply first normalization and then speckle reduction filtering for better results.

In two recent studies [54, 60], it was shown that the preprocessing of ultrasound images of the carotid artery with normalization and speckle-reduction filtering improves the performance of the automated segmentation of the intima–media thickness [54] and plaque [60]. More specifically, it was shown in Ref. [21] that a smaller variability in segmentation results was observed when performed on images after normalization and speckle-reduction filtering, compared with the manual delineation results made by two medical experts. Furthermore, in another study [7], we have shown that speckle-reduction filtering improves the percentage of correct classifications score of symptomatic and asymptomatic images of the carotid. Speckle-reduction filtering was also investigated by other researchers on ultrasound images of liver and kidney [95] and on natural scenery [14], using an adaptive two-dimensional filter similar to the *DsFlsmv* speckle-reduction filter used in this study. In these studies [14, 95], speckle-reduction filtering was evaluated based only on visual perception evaluation made by the researches.

Verhoefen et al. [96] applied mean and median filtering in simulated ultrasound images and in ultrasound images with blood vessels. The lesion-signal-to-noise ratio was used to quantify the detectability of lesions after filtering. Filtering was applied on images with fixed and adaptive size windows to investigate the influence of the filter window size. It was shown that the difference in performance between the filters was small, but the choice of the correct window size was important. Kotropoulos et al. [97] applied adaptive speckle-reduction filtering in simulated tissue mimicking phantom and liver ultrasound B-mode images, where it was shown that the proposed maximum likelihood estimator filter was superior to the mean filter.

Although in this study, speckle has been considered as noise, there are other studies where speckle, approximated by the Rayleigh distribution, was used to support automated segmentation. Specifically, in Ref. [98], an automated luminal contour segmentation method based on a statistical approach was introduced, whereas in Ref. [100], ultrasound intravascular images were segmented using knowledge-based methods. Furthermore, in Ref. [99], a semiautomatic segmentation method for intravascular ultrasound images, based on gray-scale statistics of the image was proposed, where the lumen, IMT, and the plaque were segmented in parallel by utilizing a fast marching model.

Some statistical measures, as shown in the upper part of Table 4.12, were better after normalization, and some others, shown in the bottom part of Table 4.12, were better after speckle reduction. Table 4.12 also shows that the contrast was higher for the NF and N images on both scanners and was significantly different (S) after normalization and speckle-reduction filtering (see Table 4.13). All other measures presented in Table 4.2 were comparable showing that better values were obtained on the NDS images. Moreover, it was shown that the entropy that is a measure of the information content of the image [41] was higher for both scanners in the cases of the NDS and DS images. Significantly different entropy values were obtained mostly after normalization and speckle-reduction filtering (see Table 4.13).

Low entropy images have low contrast and large areas of pixels with same or similar gray-level values. An image which is perfectly flat will have a zero entropy. On the other hand, high entropy images have high contrast and thus higher entropy values [46]. The ATL HDI-5000 scanner produces therefore images with higher information content. The entropy was also used in other studies to classify the best liver ultrasound images [44], where it was shown that the experts rated images with higher entropy values better. In Ref. [40], entropy and other texture features were used to classify between symptomatic and asymptomatic carotid plaques for assessing the risk of stroke. It was also shown [93] that asymptomatic plaques tend to be brighter, have higher entropy and are more coarse, whereas symptomatic plaques tend to be darker, have lower entropy (i.e., the image intensity in neighboring pixels is more unequal) and are less coarse. Furthermore, it is noted that texture analysis could also be performed on smaller areas of the carotid artery, such as the plaque, after segmentation [54, 60].

In previous studies [47, 48, 50, 52, 90, 91], researchers evaluated image quality on natural scenery images using either only the visual perception by experts or some of the evaluation metrics presented in Table 4.14. In this study, MSE and RMSE values were in the range of 0.4 to 2.0, for all cases, Err3, Err4, SNR, PSNR, Q, and SSIN were better between the NF–N images for both scanners, showing that normalization increases the values of these measures. In Ref. [24], speckle-reduction filtering was investigated on ultrasound images of the heart. The MSE values reported after speckle reduction for the adaptive weighted median filtering, wavelet shrinkage-enhanced filter, wavelet shrinkage filter, and nonlinear coherence diffusion were 289, 271, 132, and 121, respectively. Loupas et al. [101] applied an adaptive weighted median filter for speckle reduction in ultrasound images of the liver and gallbladder and used the speckle index and the MSE for comparing the filter with a conventional mean filter. It was shown that the filter improves the resolution of small structures in the ultrasound images. It was also documented in Ref. [47] that the MSE, RMSE, SNR, and PSNR measures are not objective for image quality evaluation and that they do not correspond to all aspects of the visual perception nor they correctly reflect artifacts [48].

Recently, the Q [50] and SSIN [47] measures for objective image quality evaluation have been proposed. The best values obtained in this study were Q=0.95 and SSIN=0.95 and were obtained for the NF–N images for both scanners. These results were followed with Q=0.73 and SSIN=0.92 in the case of NF–NDS for the HDI ATL-3000 scanner, and Q=0.72 and SSIN=0.94 in the case of NF–DS for the HDI ATL-5000 scanner. In Ref. [50], where natural scenery images were distorted by speckle noise, the values for Q reported were 0.4408, whereas the values for Q after contrast stretching were 0.9372.

The methodology presented in this study may also be applicable in future studies, to the evaluation of new ultrasound and telemedicine systems to compare their performance. It is also important to note that the methodology consists of a combination of subjective and objective measures that should be combined together for a proper image quality evaluation result [48].

5.6 VISUAL PERCEPTION AND ADDITIONAL COMMENTS BY EXPERTS

The visual perception evaluation performed in Table 4.6.a and b, showed that the filters *DsFlsmv*, *DsFgf4d*, and *DsFlsminsc* improved the visual assessment by experts. The intra-observer variability test (Table 4.6b), which was repeated 1 year after the first visual evaluation (Table 4.6a), showed that the differences between the visual evaluations made by the two experts were very low, and the results of the two tables were in agreement.

It was shown that the highest scores were obtained, for the filter *DsFlsmv* for both tables. The differences, which are observed in the ratings between the two experts, were due to the fact that each expert was interested for a different tissue area in the ultrasound image of the carotid artery. Specifically, the cardiovascular surgeon was primarily interested in the plaque composition and texture, whereas the neurovascular expert was interested in the degree of stenosis and the lumen diameter. The filter *DsFlsminsc* was rated from the neurovascular expert with the highest score in Table 4.6b. The expert found that this filter was very helpful when inspecting the degree of stenosis and the lumen diameter.

In Table 5.1, the two experts evaluated the images before and after despeckle filtering and gave some additional comments, which we think are important to be discussed briefly. It was shown that the primary interest of the experts were the borders between IMT, plaque, artery wall, and blood, to be able to exactly make a separation between them. Other important points taken into consideration from both experts during this examination were the texture of plaque, as the texture may give indication about the risk of stroke [40]. They have both commented the fact that the *DsFlsmv* filter was good for visualizing the borders between blood, plaque, and wall but not between wall and surrounding tissue, the *DsFlsminsc* helped specifically for the plaque visualization as plaque borders were better after filtering, and that the *DsFgf4d* sharpened the edges, thus it may be used for plaque visualization and to separate the borders between blood and plaque.

To the best of our knowledge, no other studies were carried out, (with the exception of Refs. [7] and [59]) on the visual evaluation of ultrasound images by using despeckle filtering and image normalization with two [65] experts. More specifically in Ref. [7], 56 different textures features and 10 different image quality evaluation metrics were used to compare the effect of despeckle filtering in 440 ultrasound images of the carotid artery, where two different experts optically evaluated the images. It was found that a linear-order statistics filter, based on first-order statistics may be successfully used for despeckling carotid artery ultrasound images. Furthermore, in Ref. [59], two different ultrasound imaging scanners, namely, the ATL HDI-3000 and the ATL HDI-5000 were compared based on texture features and image quality metrics extracted form 80 ultrasound images of the carotid bifurcation, before and after despeckle filtering. It was shown that normalization and despeckle filtering favors image quality. In a significant number of despeckle-filtering studies [5, 14, 17, 20, 24, 27, 28, 62, 63, 64, 66] visual evaluation was carried out by nonexperts. There are very few results reported in the literature, where visual perception evaluation was carried out in ultrasound images. Specifically, despeckle filtering was evaluated visually by two experts in Ref. [27], where they manually delineated 60 echocardiographic images before and after despeckle filtering. Quantitative measurements were calculated in terms of the mean of absolute border difference and the mean of border area differences. The visual evaluation in Ref. [27] showed that the borders, which were manually defined by the experts, were improved after despeckle filtering. In Ref. [78], the performance assessment of multi-temporal SAR image despeckling was evaluated from 10 photo interpreters. The evaluation was made between the original and the three filtered results. The photo interpreters evaluated the accuracy of manual detection of geographical features, such as lines points and surfaces, by presenting the images in random order. The 10 photo interpreters concluded that despeckle filtering improves the identification of the above criteria and that specific filters may be used to enhance points, lines, or surfaces as required. In another study, image quality was evaluated for compressed still images [79], where the images were presented to an unknown number of observers in random order. The observers were not experts, but they were untrained persons over 18 drawn from the university population.

5.7 SUMMARY FINDINGS ON DESPECKLE FILTERING

The results of our study showed that observer variability and sensitivity are important in image quality evaluation and can only be compensated when assessments are made against a standard scale of quality, such as the image quality evaluation metrics proposed in this study. Observer variability may also be compensated by additional tests employing image quality and texture measures, as proposed in this study, for quantifying image quality.

The findings, from the despeckle filtering, in our study may be summarized as follows (see also Table 5.1): Filter *DsFlsmv* or *DsFgf4d* can be used for despeckling asymptomatic images where the expert is interested mainly in the plaque composition and texture analysis. Filters *DsFlsmv* or *DsFgf4d* or *DsFlsminsc* can be used for despeckling of symptomatic images where the expert is interested in identifying the degree of stenosis and the plaque borders.

Filter *DsFlsmv* gave very good performance with respect to:

(a) Preserving the mean and the median, as well as decreasing the variance and the speckle index, *C*, of the image.
(b) Increasing the distance of the texture features between the asymptomatic and the symptomatic classes.

(c) Significantly changing the SGLDM range of values texture features after filtering based on the Wilcoxon rank sum test, where almost all feature sets with the exception of SAV, were significantly different.
(d) Marginally improving the classification success rate of the kNN classifier for the classification of asymptomatic and symptomatic images in the case of SF, SMF, and TEM feature sets, and
(e) Improving the image quality of the image.

The *DsFlsmv* filter, which is a simple filter, is based on local image statistics. It was first introduced in Refs. [14, 15] by Jong-Sen Lee and coworkers, and it was tested visually on a few SAR images, with satisfactory results. It was also used for SAR imaging in Refs. [9, 13] and [21] and image restoration in Ref. [11], where the evaluation was made visually by the researchers, which they have concluded that the filter showed satisfactory results.

Filter *DsFgf4d* gave very good performance with respect to:

(a) Decreasing the variance and the speckle index, *C*, and increasing the contrast significantly of the image.
(b) Marginally increasing the distance of the texture features between the asymptomatic and the symptomatic classes.
(c) Significantly changing the SGLDM range of values texture features after filtering based on the Wilcoxon rank sum test, where almost all features sets with the exception of SAV and Var were significantly different.
(d) Improving the classification success rate of the kNN classifier for the classification of asymptomatic and symptomatic images in the cases of SGLDMm, GLDS, NGTDM, SFM, and TEM feature sets.

The geometric filter *DsFgf4d* was introduced by Crimmins [10] and was tested visually on a few SAR images with satisfactory results.

Filter *DsFlsminsc* gave the best performance with respect to:

(a) Preserving the mean and the median, as well as decreasing the variance and the speckle index and increasing the contrast of the image.
(b) Increasing the distance of the texture features between the asymptomatic and the symptomatic classes.
(c) Significantly changing the SGLDM texture features after filtering based on the Wilcoxon rank sum test, where almost all features sets with the exception of the contrast, SOSV, and SAV were significantly different.

(d) Improving the classification success rate of the kNN classifier for the classification of asymptomatic and symptomatic images in the case of SF, SGLDMr, GLDS, NGTDM, FDTA and FPS feature sets.

Filter *DsFlsminsc* was originally introduced by Nagao in Ref. [32] and was tested on an artificial and an SAR image with satisfactory performance. In this study, the filter was modified, by using the speckle index instead of the variance value for each subwindow (as described in Section 2.1.3).

Filters used for speckle reduction in ultrasound imaging by other investigators include: *DsFmedian* [33], *DsFwiener* [13], *DsFhomog* [8], *DsFhomo* [17, 18], *DsFad* [5], and *DsFwaveltc* [29]. However, these filters were evaluated on a small number of images, and their performance was tested based only on the mean, median, standard deviation, and speckle index of the image before and after filtering.

The *DsFmedian* and the *DsFwiener* filters were originally used by many researchers for suppressing the additive and later for suppressing the multiplicative noise in different types of images [2–9, 13, 33]. The results of this study showed that the *DsFwiener* and *DsFmedian* filters were not able to remove the speckle noise and produced blurred edges in the filtered image (see Figures 4.1 and 4.4). In this study, the *DsFmedian* filter performed poorer as shown in Tables 4.1–4.6.

The *DsFhomog* [8] and *DsFhomo* [2, 17, 18] filters, were recently used by some researchers for speckle reduction, but our results in Tables 4.1–4.5 and the visual evaluation of the experts in Tables 4.6–4.10 showed poor performance especially for the *DsFhomo* filter.

Anisotropic diffusion is an efficient nonlinear technique for simultaneously performing contrast enhancement and noise reduction. It smoothes homogeneous image regions but retains image edges [23]. Anisotropic diffusion filters usually require many iteration steps compared with the local statistic filters. In a recent study [5], speckle-reducing anisotropic diffusion filtering was proposed as the most appropriate filter for ultrasound images of the carotid artery. However, in this study, *DsFad*, as shown in Tables 4.14–4.15 performed poorer compared to *DsFlsmv*, *DsFgf4d*, and *DsFlsminsc*.

Furthermore, wavelet filtering proposed by Donoho in Ref. [29] was investigated for suppressing the speckle noise in SAR images [15, 35], real-world images [25], and ultrasound images [26] with favorable results. In this study, it is shown that the *DsFwaveltc* filter gave poorer performance for removing the speckle noise from the ultrasound images of the carotid artery (Tables 4.1–4.3).

All the above investigators described their results as quite favorable with improvements in contrast enhancement, noise reduction, and edge preservation after filtering, but we believe that these results were not well quantified. The reason is that researchers have used a small number of images, a limited number of texture descriptors to quantify the filtered images, and they have not included findings of experts such as our findings [7, 59] from the two vascular experts. On the other hand, the methods proposed in other studies have not been evaluated and compared with other methods like in our study. For our filter evaluation, a large set of feature descriptors were employed which, when used with the statistical Wilcoxon test, the kNN-classifier, the visual assistance and the quantification of two experts offers a complete and more accurate result. In Table 4.3, the statistical test of significantly different features before and after despeckle filtering was presented, and it was shown that despeckle filters, *DsFhomo*, *DsFwaveltc*, *DsFmedian*, and *DsFwiener*, proposed by other researchers, showed bad results as far as significant difference concerns, thus the classes (asymptomatic, symptomatic) could not be easily separated. The filters *DsFlsmv*, *DsFgf4d*, and *DsFlsminsc* showed better results in this test.

Finally, Table 5.1 summarizes the findings on despeckle filtering and proposes what despeckle filter should be used if the primary interest is the plaque texture and composition or the outline of the plaque together with the degree of stenosis and the lumen diameter. Specifically, Table 5.1 suggests what filter should be used for despeckle filtering if the expert is interested in computer-aided diagnosis or visual perception evaluation. The final message is that, depending on the purpose of the ultrasound scanning or the clinical diagnosis needed from the experts, one may have to use a different despeckle filter.

While in almost all studies where despeckle filtering was investigated, statistical and or texture analysis was used to evaluate despeckle filtering, the visual perception evaluation was used in few studies, where an original image was to be evaluated with a processed or a despeckled one. Also, the number of the images used for the evaluation varied, and there is no other study reported where such a large number of images were used for evaluating despeckle filtering as in this study. It seems that visual evaluation is widely accepted in the medical community.

However, visual perception evaluation is associated with the problems [59]. As it is well known, visual perception is very subjective and suffers from intra- and inter-observer variability. Furthermore, visual analysis is subjective to both systemic and random errors. A systemic error can be introduced when changing the screen settings or observed between different experts. Although appropriate training and method standardization should eliminate these potential problems, visual perception evaluation still remains one of the biggest challenges in the automatic image analysis and evaluation. To overcome some of the difficulties of visual analysis, Haralick [41] suggested a standardization, or normalization, procedure, as explained in Ref. [59], which needs to be applied. This normalization has been also used in our study for the despeckling, segmentation, and image-quality evaluation.

* * * *

CHAPTER 6

Summary and Future Directions

6.1 SUMMARY

Despeckle filtering applications has been a rapidly emerging research area in recent years. The basic principles, the theoretical background, and the algorithmic steps of a representative set of despeckle filters were covered in this book. Moreover, selected representative applications of image despeckling covering a variety of ultrasound image processing tasks are presented. Most importantly, a despeckle filtering and evaluation protocol is documented in Table 6.1. The source code of the algorithms discussed in this book has been made available on the web, thus enabling researchers to more easily exploit the application of despeckle filtering in their problems under investigation.

A total of 11 different despeckle filters were documented in this book based on linear filtering, nonlinear filtering, diffusion filtering, and wavelet filtering. We have evaluated despeckle filtering on 440 (220 asymptomatic and 220 symptomatic) ultrasound images of the carotid artery bifurcation, based on visual evaluation by two medical experts, texture analysis measures, and image quality evaluation metrics. A linear despeckle filter based on local statistics (*DsFlsmv*) improved the class separation between the asymptomatic and the symptomatic classes, gave only a marginal improvement in the percentage of correct classifications success rate based on texture analysis and the kNN classifier, and improved the visual assessment by the experts. It was also found that the *DsFlsmv* despeckle filter can be used for despeckling asymptomatic images where the expert is interested mainly in the plaque composition and texture analysis, whereas a geometric despeckle filter (*DsFgf4d*) can be used for despeckling of symptomatic images where the expert is interested in identifying the degree of stenosis and the plaque borders. The results of this study suggest that the first-order statistics despeckle filter *DsFlsmv* may be applied on ultrasound images to improve the visual perception and automatic image analysis.

Furthermore, despeckle filtering was investigated as a preprocessing step for the automated segmentation of the IMT [54] and the carotid plaque [60], followed by the carotid plaque texture analysis, and classification (as documented in the above paragraph). Despeckle filters *DsFlsmv*, *DsFlsminsc*, and *DsFgf4d* gave the best performance for the segmentation tasks. It was shown in Ref. [54] that when normalization and speckle reduction filtering is applied on ultrasound images of the carotid artery before IMT segmentation, the automated segmentation measurements are

TABLE 6.1 Despeckle Filtering and Evaluation Protocol

DESPECKLE FILTERING AND EVALUATION PROTOCOL	
1	*Recording of ultrasound images*: Ultrasound images are acquired by ultrasound equipment and stored for further image processing. Regions of interest (ROIs) could be selected for further processing.
2	*Normalize the image*: The stored images may be retrieved, and a normalized procedure may be applied (as described for example in Section 3.2).
3	*Apply despeckle filtering*: Select the set of filters to apply despeckling together with their corresponding parameters (like moving window size, iterations, and other).
4	*Texture features analysis*: After despeckle filtering, the user may select ROIs (i.e., the plaque or the area around the IMC) and extract texture features. Distance metrics between the original and the despeckled images may be computed (as well as between different classes of images if applicable).
5	*Compute image quality evaluation metrics*: On the selected ROIs, compute image quality evaluation metrics between the original noisy and the despeckled images.
6	*Visual quality evaluation by experts*: The original and/or despeckled images may be visually evaluated by experts.
7	*Select the most appropriate despeckle filter/filters*: Based on steps 3 to 6, construct a performance evaluation table (see for example Table 5.1) and select the most appropriate filter(s) for the problem under investigation.

closer to the manual measurements. This field has also been investigated by our group [55]. Our findings showed promising results; however, further work is required to evaluate the performance of the suggested despeckle filters at a larger scale as well as their impact in clinical practice. In addition, the usefulness of the proposed despeckle filters, in portable ultrasound systems and in wireless telemedicine systems still has to be investigated.

Our results on image quality evaluation (for comparing two different ultrasound scanners, ATL HDI-3000 and ATL HDI-5000) showed that normalization and speckle-reduction filtering are important preprocessing steps favoring image quality. In addition, the usefulness of the proposed methodology based on quality evaluation metrics combined with visual evaluation in ultrasound imaging and in wireless telemedicine systems needs to be further investigated.

For those readers whose principal need is to use existing image despeckle filtering technologies and apply them on different type of images, there is no simple answer regarding which specific filtering algorithm should be selected without a significant understanding of both the filtering fundamentals and the application environment under investigation. A number of issues would need to be addressed. These include availability of the images to be processed/analyzed, the required level of filtering, the application scope (general-purpose or application-specific), the application goal (for extracting features from the image or for visual enhancement), the allowable computational complexity, the allowable implementation complexity, and the computational requirements (e.g., real-time or offline). We believe that a good understanding of the contents of this book can help the readers make the right choice in selecting the most appropriate filter for the application under development. Furthermore, the despeckle filtering evaluation protocol documented in Table 6.1 could also be exploited.

6.2 FUTURE DIRECTIONS

The despeckle filtering algorithms and the measures for image quality evaluation introduced in this book can also be generalized and applied to other image and video processing applications. Only a small number of filtering algorithms and image quality evaluation metrics were investigated in this book, and numerous extensions and improvements can be envisaged.

In general, the development of despeckle filtering algorithms for image despeckling, is a well-investigated field, and many researchers have been involved in this subject, but there is still not an appropriate method proposed, which will satisfy both the visual and the automated interpretation of image processing and analysis tasks. Most importantly, more comparative studies of despeckle filtering are necessary, where different filters could be evaluated by multiple experts as well as based on image quality and evaluation metrics as also proposed in this book.

In addition, the issue of video despeckling is still in its infancy, although it is noted that the proposed methodology and filtering algorithms documented in this book may be also investigated in video sequences (by frame filtering). There are many issues related to video despeckle filtering that remain to be solved. In general, the development of a multiplicative model based on video sequences is required, as most of the models developed for video filtering were for additive noise [120–126]. Furthermore, the utilization of the motion-detail by using motion estimation, in order to estimate pixels that need to be filtered in the neighboring frames should also be utilized as also proposed in [129].

Despeckle filtering may be also applied in the preprocessing of ultrasound images for other organs, including the detection of hyperechoic or hypoechoic lesions in the kidney, liver, spleen, thyroid, kidney, echogardiographic images, mammography, and others. It may be particularly effective when combined with harmonic imaging, as both can increase tissue contrast. Speckle reduction can also be extremely valuable when attempting to fuse ultrasound with Computed Tomography (CT), MRI, Positron Emission Tomography (PET), or Optical Coherence Tomography (OCT) images. For example, when a lesion is suspected on a CT scan but is not clearly visible, ultrasound despeckle filtering can be applied to accentuate subtle borders that may be masked by speckle.

Ultrasound imaging instrumentation, linked with imaging hardware and software technology have been rapidly advancing in the last two decades. Although these advanced imaging devices produce higher quality images and video, the need still exists for better image and video processing techniques including despeckle filtering. Towards this direction, it is anticipated that the effective use of despeckle filtering (by exploiting the filters and algorithms documented in this book) will greatly help in producing images with higher quality. These images would not be only easier to visualize and to extract useful information, but would also enable the development of more robust image preprocessing and segmentation algorithms, minimizing routine manual image analysis and facilitating more accurate automated measurements of both industrially and clinically relevant parameters.

Appendices

Appendix A.1 contains a listing of all the functions included in the despeckle filtering toolbox, as introduced in Chapter 2 of this book. It also includes all the functions used for the image quality and evaluation toolbox. Appendix A.2 presents an example in MATLAB code for a complete application of despeckling, image quality evaluation, and texture analysis. The toolbox including all the MATLAB code for despeckling, texture analysis and the image quality evaluation can be downloaded at: http://www.medinfo.cs.ucy.ac.cy.

All page numbers listed refer to pages in the book, indicating where a function is first used and illustrated.

APPENDIX A.1: Despeckle filtering, texture analysis, and quality evaluation toolbox functions

The following MATLAB functions are grouped in categories as presented in Table 1.1 of this book.

FUNCTION CATEGORY AND NAME	DESCRIPTION	PAGE OR OTHER LOCATION
	Linear filtering	
DsFlsmv	Mean and variance local statistics despeckle filter	p. 22, Algorithm 2.1
DsFlsmv1d	Minimum variance homogeneous 1D mask despeckle filter	p. 27
DsFlsmvsk2d	Mean variance, higher moments local statistics despeckle filter	p. 28, Algorithm 2.3
DsFlsminsc	Minimum speckle index homogeneous mask despeckle filter	p. 29, Algorithm 2.4
DsFwiener	Wiener despeckle filter	p. 26, Algorithm 2.2
	Nonlinear filtering	
DsFmedian	Median despeckle filter	p. 35, Algorithm 2.5
DsFls	Linear scaling of the gray-level values despeckle filter	p. 36
DsFca	Linear scaling of the gray-level despeckle filter	p. 36, Algorithm 2.6

DsFlecasort	Linear scaling and sorting despeckle filter	p. 36
DsFgf4d	Geometric despeckle filtering	p. 38, Algorithm 2.8
DsFhomog	Most homogeneous neighborhood despeckle filter	p. 37, Algorithm 2.7
DsFhomo	Homomorphic despeckle filtering	p. 44, Algorithm 2.9
	Diffusion filtering	
DsFad	Perona and Malik anisotropic diffusion filter	p. 45
DsFsrad	Speckle reducing anisotropic diffusion filter	p. 46, Algorithm 2.10
DsFnldif	Nonlinear coherent diffusion despeckle filter	p. 50, Algorithm 2.11
	Wavelet filtering	
DsFwaveltc	Wavelet despeckle filtering	p. 52, Algorithm 2.12

The following texture analysis MATLAB functions (also presented in Section 3.5, page 60 of this book) are here below described:

FUNCTION CATEGORY AND NAME	DESCRIPTION	PAGE OR OTHER LOCATION
DsTnwfos	First-order statistics (FOS) (features 1–5)	p. 60, Web site
	Texture analysis functions	
DsTnwsgldm	Haralick spatial gray-level dependence matrices (SGLDM) (6–31)	p. 60, Web site
DsTnwgldmc	Gray-level difference statistics (GLDS) (32–35)	p. 60, Web site
DsTnwngtdmn	Neighborhood gray-tone difference matrix (NGTDM) (36–40)	p. 60, Web site
DsTnwsfm	Statistical feature matrix (SFM) (41–44)	p. 60, Web site
DsTnlaws	Laws texture energy measures (TEM) (45–50)	p. 60, Web site
DsTfdta2	Fractal dimension texture analysis (FDTA) (51–54)	p. 60, Web site

DsTfps	Fourier power spectrum (FPS) (55–56)	p. 60, Web site
DsTfshape2	Shape (x, y, area, perimeter, perimeter^2/area) (57–61)	Web site
DsTintens2	Intensity difference vector with steps	Web site
DsTleast	Estimation of the curve slope using least squares	Web site
DsTresol2	Multiple resolution feature extraction	Web site
DsTexfeat	Main texture analysis function	Web site

The following image quality evaluation MATLAB functions are given as presented in Section 3.9.

FUNCTION CATEGORY AND NAME	DESCRIPTION	PAGE OR OTHER LOCATION
	Quality evaluation	
DsQEgae	Geometric average error	p. 64
DsQEmse	Mean square error	p. 64
DsQEsnr	Signal-to-noise radio	p. 65
DsQErmse	Randomized mean square error	p. 64
DsQEpsnr	Peak signal-to-noise radio	p. 65
DsQEminkowski	Minkowski metrics, 3rd (M3) and 4th (M4) moments	p. 64
DsQEimg_qi	Universal quality index	p. 66
DsQEssim_index	Structural similarity index	p. 66
DsQEget_dir_files	Get directory files	Web site
DsQE_quality_evaluation	Main quality evaluation program	Web site
DsQmetrics	Function for running all above quality evaluation metrics	Web site

APPENDIX A.2: Examples of Running the Despeckle filtering Toolbox Functions

MATLAB™ code for the *DsQmetrics.m* function

```
function f=DsQmetrics(I,K);
% I: Original input noisy image
% K: Despeckled input Image
% f: Matrix with image quality metrics

I=double(I); K=double(K);
 gaer = DsQ_gae(I,K);
 metrics= [gaer];

 % calculate the mean square error mse
 mser=DsQmse(I,K);
 metrics=[metrics, mser];

 % calculate the signal-to-noise radio snr
 snrad=DsQsnr(I,K);
 metrics=[metrics, snrad];

 % calculate the square root of the mean square error
 rmser=DsQrmse(I, K);
 metrics=[metrics, rmser];

 % Calculate the peak-signal-to-noise radio
 psnrad=DsQ_psnr(I,K);
 metrics=[metrics, psnrad];

 % Calculate the Minkwofski measure
 [M3, M4] = DsQminkowski(I, K);
 metrics=[metrics, M3, M4];

 % Calculate the universal quality index
 [quality, quality_map] = DsQimg_qi(I,K);
 metrics=[metrics, quality];
```

```
% Calculate the structural similarity index
[mssim, ssim_map] = DsQssim_index(I, K);
metrics=[metrics, mssim];

% calculate aditional metrics
[MSE,PSNR,AD,SC,NK,MD,LMSE,NAE,PQS]= DsQiq_measures(I,K);
metrics=[metrics, AD, SC, NK, MD, LMSE, NAE, PQS];
f= metrics;
```

The following code sequence will read an image and apply the *DsFlsmv* despeckle filter on the image iteratively five times, by using a moving sliding window of [7×7] pixels. The texture features as well as the image quality metrics between the original and the despeckled images are calculated with the code in Algorithm 2.1, and stored in the variable matrix A and B, respectively (see also Chapter 3.5–3.9). The image quality metrics between the original and the despeckled images are stored in the matrix M.

```
% Read the image original.tif and store it in variable image
image = imread ('original.tif');

% Apply the despeckle filter DsFlsmv on the image using a sliding moving window of 7x7 pixel,
%iteratively 5 times
despeckle = DsFlsmv (image, [7 7], 5);
% Show the original and the despeckled images on the screen
figure, imshow (image); figure, imshow (despeckle);

% Calculate the texture features for the original and the despeckled images
Orig_textfeat = DsTexfeat (image);
Desp_textfeat = DsTexFeat (despeckle);

% Save the extracted features of the original and despeckled images in the mat files A and B
save Orig_textfeat A;
save Desp_textfeat B;

% Calculate the image quality evaluation metrics between the original and the despeckled images
% and save them in a matrix M
M = DsQmetrics (image, despeckle);
```

```
%The mat files A, B can then be loaded into the MATLAB workspace for opening, reading and storing the features and image quality metrics. This can be made by double clicking the mat files.
% The command whos will show the files loaded
whos
% The open command will then open the file A, B and M
open A;
open B;
open M;
%The texture features for both the original and despeckled images and the quality evaluation
%metrics can now be manipulated or saved elsewhere.
```

List of Symbols

$a_{i,j}$	Additive noise component on pixel i,j
$_{\text{comp}},\ _{\text{comp}}$	Logarithmic compression parameters
(s)	Snake stiffness of the energy functional
$_{\text{GVF}}$	GVF snake rigidity parameter
C	Speckle index
CV%	Coefficient of variation
$cd(\quad g\ ), c_{i,j}$	Diffusion coefficient
c_{adsr}	Speckle reducing anisotropic diffusion coefficient
c	Constant controlling the magnitude of the potential
$c_{s\sin\_1}, c_{s\sin\_2}$	Constants used to calculate the SSIN
c_2	Positive weighting factor
	Number of directions, which diffusion is computed
	Signal-to-noise radio (SNR)
$D \quad ^{2x2}$	Symmetric positive semi-definite diffusion tensor representing the required diffusion in both gradient and contour directions
D_{f}	Fractal dimension
D	Matrix used to calculated the image energy of the snake, $E_{\text{image}}(\ )$
D_{viewing}	Viewing distance
DR	Dynamic range of input ultrasound signal
$d(k)$	Wavelet coefficient for the wavelet filtering
f	Frequency shift (Doppler frequency shift)
r	Distance between two pixels
g	The gradient magnitude of image $g(x,y)$(gradient)
$g_{i,j}$	Directional derivative (simple difference) at location i,j
$f_1 \cdots f_{13}$	SGLDM texture measures from Haralick
$f_x(x,y)$	First-order differential of the edge magnitude along the x-axis
$f_{i,j}$	Noise-free signal ultrasound signal in discrete form (the new image) on pixel i,j
f	Frequency of ultrasound wave
f_0	Transmitted frequency of ultrasound signal

$feat\_dis_i$	Percentage distance
$g_{i,j}$	Observed ultrasound signal in discrete formulation after logarithmic compression
$g(x,y)$	Observed ultrasound signal after logarithmic compression, representing image intensity at location (x, y)
G	Linear gain of the amplifier
$G \quad g_{i,j}$	Image convolved with Gaussian smoothing filter
G	Gaussian smoothing filter
$\bar{g}_i, \bar{f}_i$	Mean gravity of the searching pixel region in image g or f
g_{max} and g_{min}	Maximum and minimum gray-level values in a pixel neighborhood, respectively
Ha, kHz, and MHz	Hertz, kilohertz, and megahertz, respectively
HX, HY	Entropies of p_x and p_y
$H^{(k)}$	Hurst coefficients
$H(x,y)$	Array of points of the same size for the HT
HD	Hausdorff distance
$_s$	Spatial neighborhood of pixel i, j
$\lvert \ _s \rvert$	Number of neighbors (usually four except at the image boundaries)
$_i$	Phase shift relative to the insonated ultrasound wave
	Angle between the direction of movement of the moving object and the ultrasound beam
I	Identity matrix
$I_0(x)$	Modified Bessel function of the first kind of order 0
$I_1 \quad I_7$	Echo boundaries describing the regions in carotid artery
IMT_{mean}	Mean value of the IMT
IMT_{min}	IMT minimum value
IMT_{max}	IMT maximum value
IMT_{median}	IMT median value
k	Coefficient of variation for speckle filtering
	Wavelength of ultrasound wave
	Lai and Chin snake energy regularization parameter, $E_{snake}(\ )$
$_d \quad ^+$	Rate of diffusion for the anisotropic diffusion filter
m_{i1} and m_{i2}	Mean values of two classes (asymptomatic and symptomatic, respectively)

m/s and cm/s	Meters per second and centimeters per second, respectively
	Mean
N	Number of scatterers within a resolution cell
N_{feat}	Number of features in the feature set
$n_{i,j}$	Multiplicative noise component (independent of $g_{i,j}$, with mean 0) on pixel i, j
$nl_{i,j}$	Multiplicative noise component after logarithmic compression on pixel i, j
$n(s)$	Normal force tensor
$_i$	Amount of ultrasound signal backscattered by scatterer i
$p_x(i)$	ith entry in the marginal probability matrix obtained by summing the rows of $p(i, j)$
Q	Mathematically defined universal quality index
$R = 1 \quad \frac{1}{1 + \ ^2}$	Smoothness of an image
$Score\_Dis$	Score distance between two classes (asymptomatic, symptomatic)
$s_e = \ _{IMT} / \ \overline{2}$	Inter-observer error
s_{max}	Maximum pixel value in the image
s^2	Structural energy
$_{\text{IMT}}$	IMT standard deviation
$_{fg}$	Covariance between two images f and g
	Standard deviation
^{2}	Variance
$_3$	Skewness
$_4$	Kurtosis
$_{i1}$ and $_{i2}$	Standard deviations of two classes (asymptomatic and symptomatic, respectively)
$2 \ ^2$	Diffuse energy
$_n$	Standard deviation of the noise
^2_w	Variance of the gray values in a pixel window

List of Abbreviations

ACSRS	Asymptomatic carotid stenosis
DsFad	Perona and Malik anisotropic diffusion filter
DsFadsr	Speckle reducing anisotropic diffusion filter
ASM	Angular second moment
ATL HDI-3000	ATL 3000 ultrasound scanner
ATL HDI-5000	ATL 5000 ultrasound scanner
DsFca	Linear scaling of the gray-level despeckle filter
CAT	Computer-assisted tomography
CCA	Common carotid artery
CSR	Contrast-to-speckle radio
CT	Computer tomography
CW	Continuous wave
DR	Dynamic range
DS	Despeckled
DSCQS	Double stimulus continuous quality scale
DSIS	Double stimulus impairment scale
DVD	Digital video
DWT	Discrete wavelet transform
E	Effectiveness measure
ECA	External carotid artery
ECST	European carotid surgery trial
DsQEErr	Error summation in the form of the Minkowski metric
FDTA	Fractal dimension texture analysis
FFT	Fast Fourier transform
FPS	Fourier power spectrum
GAE	Geometric average error
GF	Geometric filtering
DsFgf4d	Geometric despeckle filter
DsFgfminmax	Geometric despeckle filter utilizing minimum maximum values
GGVF	Generalized gradient vector flow

GLDS	Gray-level difference statistics
GVF	Gradient vector flow
HD	Hausdorff distance
HF	Maximum homogeneity
HM	Homomorphic
DsFhomo	Homomorphic despeckle filter
DsFhomog	Most homogeneous neighbourhood despeckle filter
HVS	Human visual system
ICA	Internal carotid artery
IDM	Inverse difference moment
IDV	Intensity difference vector
IMC	Intima–media complex
IMT	Intima–media thickness
IVUS	Intravascular ultrasound
kNN	The statistical *k*-nearest-neighbor classifier
DsFlecasort	Linear scaling and sorting despeckle filter
LS	Linear scaling
DsFls	Linear scaling of the gray-level values despeckle filter
DsFlsmedcd	Lee diffusion despeckle filter
DsFlsminsc	Minimum speckle index homogeneous mask despeckle filter
DsFlsminv1d	Minimum variance homogeneous 1D mask despeckle filter
DsFlsmv	Mean and variance local statistics despeckle filter
DsFlsmvsk2d	Mean variance, higher moments local statistics despeckle filter
DsFlsmvske1d	Mean, variance, skewness, kurtosis 1D local statistics despeckle filter
M	Manual
DsFmedian	Median despeckle filter
MF	Multi-resolution fractal
MMSE	Minimum mean-square error
MN	Manual normalized
MRI	Magnetic resonance imaging
MSE	Mean square error
N	Normalized
ND	Normalized despeckled
NE	North–east
NF	No filtering

NGTDM	Neighborhood gray-tone difference matrix
NIE	Normalized image energy
DsFnldif	Nonlinear coherent diffusion despeckle filter
NS	Not significant difference
NST	North–south
NTSE	Normalized total snake energy
P	Precision
PDE	Partial differential equation
PDF	Probability density function
PET	Positron emission tomography
DsQEPSNR	Peak signal-to-noise radio
PW	Pulsed wave
R	Sensitivity (or recall)
RF	Radio frequency
DsQERMSE	Root mean square error
ROC	Receiver operating characteristic
S	Significant difference
Sp	Specificity
SAR	Synthetic aperture radar
SD	Standard deviation
SE	South–east
SFM	Statistical feature matrix
SGLDM	Spatial gray-level dependence matrices
SGLDMm	Spatial gray-level dependence matrix mean values
SGLDMr	Spatial gray-level dependence matrix range of values
DsQESNR	Signal-to-noise radio
SPECT	Single photon emission computer tomography
SSIN	Structural similarity index
TEM	Laws texture energy measures
TGC	Time gain compensation
TIA	Transient ischemic attacks
TV	Television
DsFwaveltc	Wavelet despeckle filter
WE	West–east
wiener	Wiener despeckle filter

WN	West–north
WS	West–south
WT	Wavelet transform
$_{err}$	Minkowski error coefficient
1D	One-dimensional
2D	Two-dimensional
3D	Three-dimensional

References

[1] D. Lamont, L. Parker, M. White, N. Unwin, S.M.A. Bennett, M. Cohen, D. Richardson, H.O. Dickinson, A. Adamson, K.G.M.M. Alberti, A.W. Craft, "Risk of cardiovascular disease measured by carotid intima–media thickness at age 49–51: Life course study," *BMJ*, vol. 320, pp. 273–278, January 29, 2000.

[2] C.B. Burckhardt, "Speckle in ultrasound B-mode scans," *IEEE Trans. Sonics Ultrasonics*, vol. SU-25, no. 1, pp. 1–6, 1978.

[3] R.F. Wagner, S.W. Smith, J.M. Sandrik, and H. Lopez, "Statistics of speckle in ultrasound B-scans," *IEEE Trans. Sonics Ultrasonics*, vol. 30, pp. 156–163, 1983.

[4] J.W. Goodman, "Some fundamental properties of speckle," *J. Opt. Soc. Am.*, vol. 66, no. 11, pp. 1145–1149, 1976.

[5] Y. Yongjian and S.T. Acton, "Speckle reducing anisotropic diffusion," *IEEE Trans. Image Process.*, vol. 11, no. 11, pp. 1260–1270, November 2002.

[6] R.W. Prager, A.H. Gee, G.M. Treece, and L. Berman, "Speckle detection in ultrasound images using first order statistics," *GUED/F-INFENG/TR 415*, University of Cambridge, Dept. of Engineering, pp. 1–17, July 2002.

[7] C.P. Loizou, C.S. Pattichis, C.I. Christodoulou, R.S.H. Istepanian, M. Pantziaris, and A. Nicolaides "Comparative evaluation of despeckle filtering in ultrasound imaging of the carotid artery," *IEEE Trans. Ultrason. Ferroelectr. Freq. Control*, vol. 52, no. 10, pp. 1653–1669, 2005.

[8] C.I. Christodoulou, C. Loizou, C.S. Pattichis, M. Pantziaris, E. Kyriakou, M.S. Pattichis, C.N. Schizas, and A. Nicolaides,"Despeckle filtering in ultrasound imaging of the carotid artery," in *Second Joint EMBS/BMES Conference, Houston, TX*, vol. 2, pp. 1027–1028, October 23–26, 2002.

[9] J.S. Lee, "Speckle analysis and smoothing of synthetic aperture radar images," *Comp. Graphics Image Process.*, vol. 17, pp. 24–32, 1981.

[10] L. Busse, T.R. Crimmins, and J.R. Fienup, "A model based approach to improve the performance of the geometric filtering speckle reduction algorithm," *IEEE Ultrasonic Symp.*, vol. 2, pp. 1353–1356, 1995.

[11] D.T. Kuan, A.A. Sawchuk, T.C. Strand, and P. Chavel, "Adaptive restoration of images with speckle," *IEEE Trans. Acoust.*, vol. ASSP-35, pp. 373–383, 1987.

[12] M. Insana, T.J. Hall, G.C. Glendon, and S.J. Posental, "Progress in quantitative ultrasonic imaging," *SPIE Vol. 1090 Medical Imaging III, Image Formation*, pp. 2–9, 1989.

[13] V.S. Frost, J.A. Stiles, K.S. Shanmungan, and J.C. Holtzman, "A model for radar images and its application for adaptive digital filtering of multiplicative noise," *IEEE Trans. Pattern Anal. Mach. Intell.*, vol. 4, no. 2, pp. 157–165, 1982.

[14] J.S. Lee, "Digital image enhancement and noise filtering by using local statistics," *IEEE Trans. Pattern Anal. Mach. Intell.*, PAMI-2, no. 2, pp. 165–168, 1980.

[15] J.S. Lee, "Refined filtering of image noise using local statistics," *Comput. Graphics Image Process*, vol. 15, pp. 380–389, 1981.

[16] D.T. Kuan and A.A. Sawchuk, "Adaptive noise smoothing filter for images with signal dependent noise," *IEEE Trans. Pattern Anal. Mach. Intell.*, vol. PAMI-7, no. 2, pp. 165–177, 1985.

[17] S. Solbo and T. Eltoft, "Homomorphic wavelet based-statistical despeckling of SAR images," *IEEE Trans. Geosci. Remote Sens.*, vol. 42, no. 4, pp. 711–721, 2004.

[18] J. Saniie, T. Wang, and N. Bilgutay, "Analysis of homomorphic processing for ultrasonic grain signal characterization," *IEEE Trans. Ultrason. Ferroelectr. Freq. Control*, vol. 3, pp. 365–375, 1989.

[19] S. Jin, Y. Wang, and J. Hiller, "An adaptive non-linear diffusion algorithm for filtering medical images," *IEEE Trans. Inf. Technol. Biomed.*, vol. 4, no. 4, pp. 298–305, December 2000.

[20] J. Weickert, B. Romery, and M. Viergever, "Efficient and reliable schemes for nonlinear diffusion filtering," *IEEE Trans. Image Process.*, vol. 7, pp. 398–410, 1998.

[21] N. Rougon and F. Preteux, "Controlled anisotropic diffusion," in *Conference on Nonlinear Image Processing VI, IS&T/SPIE Symposium on Electronic Imaging, Science and Technology, San Jose, CA*, pp. 1–12, February 5–10, 1995.

[22] M. Black, G. Sapiro, D. Marimont, and D. Heeger, "Robust anisotropic diffusion," *IEEE Trans. Image Process.*, vol. 7, no. 3, pp. 421–432, March 1998.

[23] P. Rerona and J. Malik, "Scale-space and edge detection using anisotropic diffusion," *IEEE Trans. Pattern Anal. Mach. Intell.*, vol. 12, no. 7, pp. 629–639, July 1990.

[24] K. Abd-Elmoniem, A.-B. Youssef, and Y. Kadah, "Real-time speckle reduction and coherence enhancement in ultrasound imaging via nonlinear anisotropic diffusion," *IEEE Trans. Biomed. Eng.*, vol. 49, no. 9, pp. 997–1014, Sept. 2002.

[25] S. Zhong and V. Cherkassky, "Image denoising using wavelet thresholding and model selection," in *Proceedings of the IEEE International Conference on Image Processing, Vancouver, BC, Canada*, pp. 1–4, November 2000.

[26] A. Achim, A. Bezerianos and P. Tsakalides, "Novel Bayesian multiscale method for speckle removal in medical ultrasound images," *IEEE Trans. Med. Imaging*, vol. 20, no. 8, pp. 772–783, 2001.

[27] X. Zong, A. Laine, and E. Geiser, "Speckle reduction and contrast enhancement of echocardiograms via multiscale nonlinear processing," *IEEE Trans. Med. Imaging*, vol. 17, no. 4, pp. 532–540, 1998.

[28] X. Hao, S. Gao, and X. Gao, "A novel multiscale nonlinear thresholding method for ultrasonic speckle suppressing," *IEEE Trans. Med. Imaging*, vol. 18, no. 9, pp. 787–794, 1999.

[29] D.L. Donoho, "Denoising by soft thresholding," *IEEE Trans. Inform. Theory*, vol. 41, pp. 613–627, 1995.

[30] A.M. Wink and J.B.T.M. Roerdink, "Denoising functional MR images: A comparison of wavelet denoising and Gaussian smoothing," *IEEE Trans. Med. Imaging*, vol. 23, no. 3, pp. 374–387, 2004.

[31] V. Dutt, "Statistical analysis of ultrasound echo envelope," Ph.D. dissertation, Mayo Graduate School, Rochester, MN, 1995.

[32] M. Nagao and T. Matsuyama, "Edge preserving smoothing," *Computer Graphic and Image Processing*, vol. 9, pp. 394–407, 1979.

[33] T. Huang, G. Yang, and G. Tang, "A fast two-dimensional median filtering algorithm," *IEEE Trans. Acoustics, Speech and Signal Processing*, vol. 27, no. 1, pp. 13–18, 1979.

[34] S.M. Ali and R.E. Burge, "New automatic techniques for smoothing and segmenting SAR images," *Signal Processing*, North-Holland, vol. 14, pp. 335–346, 1988.

[35] F.N.S Medeiros, N.D.A. Mascarenhas, R.C.P Marques, and C.M. Laprano, "Edge preserving wavelet speckle filtering," in *5th IEEE Southwest Symposium on Image Analysis and Interpretation, Santa Fe, NM*, pp. 281–285, April 7–9, 2002.

[36] P. Moulin, "Multiscale image decomposition and wavelets," in *Handbook of Image & Video Processing*, A. Bovik (ed.), Academic Press, New York, pp. 289–300, 2000.

[37] P. Scheunders, "Wavelet thresholding of multivalued images," *IEEE Trans. Image Process.*, vol. 13, no. 4, pp. 475–483, 2004.

[38] S. Gupta, R.C. Chauhan, and S.C. Sexana, "Wavelet-based statistical approach for speckle reduction in medical ultrasound images," *Med. Biol. Eng. Comput.*, vol. 42, pp. 189–192, 2004.

[39] Philips Medical System Company, "Comparison of image clarity, SonoCT real-time compound imaging versus conventional 2D ultrasound imaging," *ATL Ultrasound Report*, 2001.

[40] C. Christodoulou, C. Pattichis, M. Pantziaris, and A. Nicolaides, "Texture-based classification of atherosclerotic carotid plaques," *IEEE Trans. Med. Imaging*, vol. 22, no. 7, pp. 902–912, 2003.

[41] R.M. Haralick, K. Shanmugam, and I. Dinstein, "Texture features for image classification," *IEEE Trans. Syst. Man Cybern.*, vol. SMC-3, pp. 610–621, November 1973.

[42] J.S Weszka., C.R. Dyer, and A. Rosenfield, "A comparative study of texture measures for terrain classification," *IEEE Trans. Syst. Man Cybern.*, vol. SMC-6, pp. 269–285, April 1976.

[43] M. Amadasun and R. King, "Textural features corresponding to textural properties," *IEEE Trans. Syst. Man Cybern.*, vol. 19, no. 5, pp. 1264–1274, Sept. 1989.

[44] C.M. Wu, Y.C. Chen, and K.-S. Hsieh, "Texture features for classification of ultrasonic images," *IEEE Trans. Med. Imaging*, vol. 11, pp. 141–152, June 1992.

[45] T.J. Chen, K.S. Chuang, Jay Wu, S.C. Chen, I.M. Hwang, and M.L. Jan, "A novel image quality index using Moran I statistics," *Phys. Med. Biol.*, vol. 48, pp. 131–137, 2003.

[46] R. Gonzalez and R. Woods, *Digital image processing,* 2nd ed., Addison-Wesley Longman Publishing Co., Inc. Boston, MA, USA, 2002.

[47] Z. Wang, A. Bovik, H. Sheikh, and E. Simoncelli, "Image quality assessment: From error measurement to structural similarity," *IEEE Trans. Image Process.*, vol. 13, no. 4, pp. 600–612, April 2004.

[48] S. Winkler, "Vision models and quality metrics for image processing applications," PhD Dissertation, University of Lausanne, Lausanne, Switzerland, December 21, 2000.

[49] D. Sakrison, "On the role of observer and a distortion measure in image transmission," *IEEE Trans. Comm.*, vol. 25, pp. 1251–1267, November 1977.

[50] Z. Wang and A. Bovik, "A universal quality index," *IEEE Signal Process. Lett.*, vol. 9, no. 3, pp. 81–84, March 2002.

[51] H.R. Sheikh, A.C. Bovik, and G. de Veciana, "An information fidelity criterion for image quality assessment using natural scene statistics," *IEEE Trans. Image Process.*, vol. 14, no. 12, pp. 2117–2128, December 2005.

[52] E. Krupinski, H. Kundel, P. Judy, and C. Nodine, "The medical image perception society, key issues for image perception research," *Radiology*, vol. 209, pp. 611–612, 1998.

[53] T. Elatrozy, A. Nicolaides, T. Tegos, A. Zarka, M. Griffin, and M. Sabetai, "The effect of B-mode ultrasonic image standardization of the echodensity of symptomatic and asymptomatic carotid bifurcation plaque," *Int. Angiol.*, vol. 17, pp. 179–186, no. 3, Sept. 1998.

[54] C.P. Loizou, C.S. Pattichis, M. Pantziaris, T. Tyllis, and A. Nicolaides, "Snakes based segmentation of the common carotid artery intima media," *Med. Biol. Eng. Comput.*, vol.45, no. 1, pp. 35–49, January 2007.

[55] C.S. Pattichis, C. Christodoulou, E. Kyriakou, M. Pantziaris, A. Nicolaides, M.S. Pattichis, and C.P. Loizou, "Ultrasound imaging of carotid atherosclerosis," in *Wiley encyclopaedia of biomedical engineering*, M. Akay (ed.), Hoboken, NJ: John Wiley & Sons, Inc., 2006.

[56] J.C. Dainty, *Laser speckle and related phenomena*, Berlin: Springer-Verlag, 1974.

[57] J.U. Quistgaard, "Signal acquisition and processing in medical diagnostic ultrasound," *IEEE Signal Process. Mag.*, vol. 14, no. 1, pp. 67–74, January 1997.

[58] H. Paul and H.P. Schwann, "Mechanism of absorption of ultrasound in liver tissue," *J Acoust. Soc. Am.*, vol. 50, pp. 692, 1971.

[59] C.P. Loizou, C.S. Pattichis, M. Pantziaris, T. Tyllis, and A. Nicolaides, "Quantitative quality evaluation of ultrasound imaging in the carotid artery," *Med. Biol. Eng. Comput.*, vol. 44, no. 5, pp. 414–426, 2006.

[60] C.P. Loizou, C.S. Pattichis, M. Pantziaris, and A. Nicolaides, "An integrated system for the segmentation of atherosclerotic carotid plaque," *IEEE Trans. Inform. Technol. Biomed.*, vol. 11, no. 5, pp. 661–667, November 2007.

[61] T. Greiner, C.P. Loizou, M. Pandit, J. Mauruschat, and F.W. Albert, "Speckle reduction in ultrasonic imaging for medical applications," *Proceedings of the ICASSP91, 1991 International Conference on Acoustic Signal Speech and Processing, Toronto, ON, Canada*, pp. 2993–2996, May 14–17, 1991.

[62] R.N. Czerwinski, D.L. Jones, and W.D. O'Brien, "Detection and boundaries in speckle images—Application to medical ultrasound," *IEEE Trans. Med. Imaging*, vol. 18, no. 2, pp. 126–136, February 1999.

[63] M. Karaman, M. Alper Kutay, and G. Bozdagi, "An adaptive speckle suppression filter for medical ultrasonic imaging," *IEEE Trans. Med. Imaging*, vol. 14, no. 2, pp. 283–292, 1995.

[64] A.M. Wink and J.B.T.M. Roerdink, "Denoising functional MR images: A comparison of wavelet denoising and Gaussian smoothing," *IEEE Trans. Med. Imaging*, vol. 23, no. 3, pp. 374–387, 2004.

[65] S. Jin, Y. Wang, and J. Hiller, "An adaptive non-linear diffusion algorithm for filtering medical images," *IEEE Trans. Inform. Technol. Biomed.*, vol. 4, no. 4, pp. 298–305, December 2000.

[66] H.-L. Eng and K.-K. Ma, "Noise adaptive soft-switching median filter," *IEEE Trans. Image Process.*, vol. 10, no. 2, pp. 242–251, 2001.

[67] B. Fetics, E.Y. Wong, T. Murabarashi, G.S. Nelson, M.-M. Cohen, C.E. Rochitte, J.L. Weiss, D.A. Kass, E. Nevo, "Enhancement of contrast echocardiography by image variability analysis," *IEEE Trans. Med. Imaging*, vol. 20, no. 11, pp. 1123–1130, November 2001.

[68] F. Rakebrandt, "Relation between ultrasound texture classification images and histology of atherosclerotic plaque," *Ultrasound Med. Biol.*, vol. 26, no. 9, pp. 1393–1402, 2000.

[69] M. Pattichis, C. Pattichis, M. Avraam, A. Bovik, and K. Kyriakou, "AM–FM texture segmentation in electron microscopic muscle imaging," *IEEE Trans. Med. Imaging*, vol. 19, no. 12, pp. 1253–1258, 2000.

[70] N. Mudigonda, R. Rangayyan, and J. Desautels, "Detection of breast masses in mammograms by density slicing and texture flow-field analysis," *IEEE Trans. Med. Imaging*, vol. 20, no. 12, pp. 121–1227, December 2001.

[71] C.I. Christodoulou, S.C. Michaelides, and C.S. Pattichis, "Multi-feature texture analysis for the classification of clouds in satellite imagery," *IEEE Trans. Geosci. Remote Sens.*, vol. 41, no. 11, pp. 2662–2668, November 2003.

[72] C.K. Zarins, C.Xu, and S. Glagov, "Atherosclerotic enlargement of the human abdominal aorta," *Atherosclerosis*, vol. 155, no. 1, pp. 157–164, 2001.

[73] N. Mudigonda, R. Rangayyan, and J. Desautels, "Detection of breast masses in mammograms by density slicing and texture flow-field analysis," *IEEE Trans. Med. Imaging*, vol. 20, no. 12, pp. 121–1227, December 2001.

[74] A. Pommert and K. Hoehne, "Evaluation of image quality in medical volume visualization: The state of the art," in *Medical image computing and computer-assisted intervention*, T. Dohi and R. Kikinis (eds.), *Proceedings of the MICCAI, 2002, Part II, Lecture Notes in Computer Science 2489*, Berlin: Springer-Verlag, pp. 598–605, 2002.

[75] J.E. Wilhjelm, M.S. Jensen, S.K. Jespersen, B. Sahl, and E. Falk, "Visual and quantitative evaluation of selected image combination schemes in ultrasound spatial compound scanning," *IEEE Trans. Med. Imaging*, vol. 23, no. 2, pp. 181–190, 2004.

[76] M. Eckert, "Perceptual quality metrics applied to still image compression," in *Canon information systems research*, Faculty of Engineering, University of Technology, Sydney, Australia, pp. 1–26, 2002.

[77] A. Baraldi and F. Pannigianni, "A refined gamma MAP SAR speckle filter with improved geometrical adaptivity," *IEEE Trans. Geosci. Remote Sens.*, vol. 33, no. 5, pp. 1245–1257, September 1995.

[78] E. Trouve, Y. Chambenoit, N. Classeau, and P. Bolon, "Statistical and operational performance assessment of multi-temporal SAR image filtering," *IEEE Trans. Geosc. Remote Sens.*, vol. 41, no. 11, pp. 2519–2539, 2003.

[79] D. Schilling and P.C. Cosman, "Image quality evaluation based on recognition times for fast browsing image applications," *IEEE Trans. Multimedia*, vol. 4, no. 3, pp. 320–331, Sept. 2002.

[80] B.B. Mandelbrot, *The fractal geometry of nature*, San Francisco, CA: Freeman, 1982.

[81] L.J. Porcello, N.G. Massey, R.B. Ines, and J.M. Marks, "Speckle reduction in synthetic aperture radar images," *J. Opt. Soc. Am.*, vol. 66, no. 11, pp. 1305–1311, November 1976.

[82] C.P. Loizou, C. Christodoulou, C.S. Pattichis, R. Istepanian, M. Pantziaris, and A. Nicolaides, "Speckle reduction in ultrasound images of atherosclerotic carotid plaque," in *DSP*

2002, Proceedings of the IEEE 14th International Conference on Digital Signal Processing, Santorini, Greece, pp. 525–528, July 1–3, 2002.

[83] J.M. Thijssen, B.J. Oosterveld, P.C. Hartman, G.J. Rosenbusch, "Correlations between acoustic and texture parameters from RF and B-mode liver echograms," *Ultrasound Med. Biol.*, vol. 19, pp. 13–20, 1993.

[84] Imaging Department, Johns Hopkins Bayview Medical Center, Baltimore, MD, http://www.jhbmc.jhu.edu/Imaging/index.html.

[85] K.T. Dussik, "On the possibility of using ultrasound waves as a diagnostic aid," *Neurol. Psychiatr.*, vol. 174, pp. 153–168, 1942.

[86] A. Kurjak, "Ultrasound scanning—Prof. Ian Donald (1910–1987)," *Eur. J. Obstet. Gynecol. Reprod. Biol.*, vol. 90, no. 2, pp. 187–189, June 2000.

[87] S.-M. Wu, Y.-W. Shau, F.-C. Chong, and F.-J. Hsieh, "Non-invasive assessment of arterial dimension waveforms using gradient-based Hough transform and power Doppler ultrasound imaging," *J. Med. Biol. Eng. Comput.*, vol. 39, pp. 627–632, 2001.

[88] W.R. Hedrick and D.L. Hykes, "Image and signal processing in diagnostic ultrasound imaging," *Journal of Diagnostic Medical Sonography*, vol. 5, no. 5, pp. 231–239, 1989.

[89] F.J. Polak, "Doppler sonography: An overview," in *Peripheral vascular sonography: A practical guide*, Baltimore, MD: Williams & Wilkins, pp. 155–160, 1992.

[90] A. Ahumada and C. Null, "Image quality: A multidimensional problem," in *Digital images and human vision*, A.B. Watson (ed.), Cambridge, MA: Bradford Press, pp. 141–148, 1993.

[91] G. Deffner, "Evaluation of display image quality: Experts vs. non-experts," *Symp. Soc. Inf. Display Dig.*, vol. 25, pp. 475–478, 1994.

[92] E.A. Fedorovskaya, H. De Ridder, and F.J. Blomaert, "Chroma variations and perceived quality of colour images and natural scenes," *Color Res. Appl.*, vol. 22, no. 2, pp. 96–110, 1997.

[93] E. Kyriakou, M.S. Pattichis, C.I. Christodoulou, C.S. Pattichis, S. Kakkos, M. Griffin, and A.N. Nicolaides, "Ultrasound imaging in the analysis of carotid plaque morphology for the assessment of stroke," in *Plaque imaging: Pixel to molecular level*, J.S. Suri, C. Yuan, D.L. Wilson, and S. Laxminarayan (eds.), IOS Press, Amsterdam, pp. 241–275, 2005.

[94] A.N. Nicolaides, M. Sabetai, S.K. Kakkos, S. Dhanjil, T. Tegos, and J.M. Stevens, "The asymptomatic carotid stenosis and risk of stroke study," *Int. Angiol.*, vol. 22, no. 3, pp. 263–272, 2003.

[95] J.C. Bamber and C. Daft, "Adaptive filtering for reduction of speckle in ultrasonic pulse-echo images," *Ultrasonic*, vol. 24, pp. 41–44, 1986.

[96] J.T.M. Verhoeven and J.M. Thijssen, "Improvement of lesion detectability by speckle reduction filtering: A quantitative study," *Ultrasonic Imaging*, vol. 15, pp. 181–204, 1993.

[97] C. Kotropoulos and I. Pitas, "Optimum nonlinear signal detection and estimation in the presence of ultrasonic speckle," *Ultrasonic Imaging*, vol. 14, pp. 249–275, 1992.

[98] E. Brusseau, C.L. De Korte, C.L., F. Mastick, J. Schaar, and A.F.W. Van der Steen, "Fully automatic luminal contour segmentation in intracoronary ultrasound imaging—A statistical approach," *IEEE Trans. Med. Imaging*, vol. 23, no. 5, pp. 554–566, 2004.

[99] M.R. Cardinal, J. Meunier, G. Soulez, E. Thérasse, and G. Cloutier, "Intravascular ultrasound image segmentation: A fast-marching method," *Proc. MICCAI, LNCS 2879*, pp. 432–439, 2003.

[100] M.E. Olszewski, A. Wahle, S.C. Vigmostad, and M. Sonka, "Multidimensional segmentation of coronary intravascular ultrasound images using knowledge-based methods," *Med. Imaging: Image Process. Proc. SPIE*, 5747, pp. 496–504, 2005.

[101] T. Loupas, W.N. McDicken, and P.L. Allan, "An adaptive weighted median filter for speckle suppression in medical ultrasonic images," *IEEE Trans. Circuits Syst.*, vol. 36, pp. 129–135, 1989.

[102] A. Zahalka and A. Fenster, "An automated segmentation method for three-dimensional carotid ultrasound images," *Phys. Med. Biol.*, vol. 46, pp. 1321–1342, 2001.

[103] A. Hamou and M. El.-Sakka, "A novel segmentation technique for carotid ultrasound images," *Int. Conf. Acoustic Speech Signal Process., ICASSP*, vol. 3, pp. III-521–III-524, 2004.

[104] D.F. Specht, "Probabilistic neural networks," *INNS Neural Networks*, vol. 3, no. 1, pp. 109–118, 1990.

[105] D. Williams and M. Shah, "A fast algorithm for active contour and curvature estimation," *GVCIP: Imag. Und.*, vol. 55, no. 1, pp. 14–26, 1992.

[106] L.D. Cohen, "On active contour models and balloons," *CVGIP:IU*, vol. 53, no. 2, pp. 211–218, 1991.

[107] K.F. Lai and R.T. Chin, "Deformable contours-modeling and extraction," *IEEE Trans. PAMI*, vol. 17, no. 11, pp. 1084–1090, 1995.

[108] C. Xu and . Prince, "Generalized Gradient vector flow external forces for active contours," *Signal Process.*, vol.71, pp. 131–139, 1998.

[109] J. Smilowitz, J. Balog, H. Keller, G. Olivera, L.A. Dewerd, and T.R. Mackie, "A new multipurpose quality assurance phantom for clinical tomotherapy," in *Proceedings of the 22nd Annual EMBS International Conference*, pp. 1191–1194, July 23–28, 2000.

[110] Loyola University Medical Center, Loyola University Chicago Stritch School of Medicine, Chicago, IL, http://www.meddean.luc.edu/lumen/MedEd/Neuro/index.htm.

[111] M.L. Bots, A.W. Hoes, P.J. Koudstaal, A. Hofman A, and D.E. Grobbee, "Common carotid intima-media thickness and risk of stroke and myocardial infarction: The Rotterdam Study," *Circulation*, vol. 96, pp. 1432–1437, 1997.

[112] J. E. Wilhjelm, M.L. Gronholdt, B. Wiebe, S.K. Jespersen, L.K. Hansen, and H. Sillesen, "Quantitative analysis of ultrasound B-mode images of carotid atherosclerotic plaque: Correlation with visual classification and histological examination," *IEEE Trans. Med Imaging*, vol. 17, no. 6, pp. 910–922, 1998.

[113] D. Lamont, L. Parker, M. White, N. Unwin, S.M.A. Bennett, M. Cohen, D. Richardson, H.O. Dickinson, A. Adamson, K.G.M.M. Alberti, A.W. Craft, "Risk of cardiovascular disease measured by carotid intima–media thickness at age 49–51: Life course study," *BMJ*, vol. 320, pp. 273–278, 2000.

[114] C.K. Zarins, C. Xu, and S. Glagov, "Atherosclerotic enlargement of the human abdominal aorta," *Atherosclerosis*, vol. 155, no. 1, pp. 157–164, 2001.

[115] ACAS Clinical Advisory, "Carotid endarterectomy for patients with asymptomatic internal carotid artery stenosis," *Stroke*, vol. 25, no. 12, pp. 2523–2524, 1994.

[116] Executive Committee for the Asymptomatic Carotid Atherosclerosis study, "Endarterectomy for asymptomatic carotid stenosis," *J. Am. Med. Assoc.*, vol. 273, pp. 1421–1428, 2002.

[117] Y. Wang, J. Ostermann, and Y.-Q. Zhang, "Video processing and communications," in *Signal processing series*, A.V. Openheim (ser. ed.), Prentice-Hall, Upper Saddle River, New Jersey, USA, pp. 1–32, 2002.

[118] C. Stller and J. Konrad, "Estimating motion in image sequences," *IEEE Signal Process. Mag.*, vol. 16, pp. 70–91, July 1999.

[119] A. Kokaram, "Motion picture restoration," in *Digital algorithms for artefact suppression in degraded motion picture film and video*, London: Springer-Verlag, ch. 10, pp. 241–259, 1998.

[120] S. Winkler, "Digital video quality," in *Vision models and metrics*, John Wiley & Sons Ltd., England, pp. 71–102, 2005.

[121] M. Oezkan, A. Erdem, M. Sezan, and A. Tekalp, "Efficient multi-frame Wiener restoration of blurred and noisy image sequences," *IEEE Trans. Image Process.*, vol. 1, pp. 453–476, October 1992.

[122] P.M.B. Van Roosmalen, S.J.P Westen, R.L. Lagendijk, and J. Biemond, "Noise reduction for image sequences using an oriented pyramid threshold technique," *IEEE Int. Conf. Image Process.*, vol. 1, pp. 375–378, 1996.

[123] M. Vetterli and J. Kovacevic, *Wavelets and subband coding*, Prentice-Hall, Signal Processing Series, Englewood Cliffs, NJ, 1995.

[124] J.-H. Jung, K. Hong, and S. Yang, "Noise reduction using variance characteristics in noisy image sequence," *Int. Conf. Consumer Electron.*, vol. 1, pp. 213–214, January 8–12, 2005.

[125] M. Bertalmio, V. Caselles, and A. Pardo, "Movie denoising by average of warped lines," *IEEE Trans. Image Process.*, vol. 16, no. 9, pp. 233–2347, 2007.

[126] B. Alp, P. Haavisto, T. Jarske, K. Oestaemoe, and Y. Neuro, "Median based algorithms for image sequence processing," *SPIE Vis. Commun. Image Process.*, pp. 122–133, 1990.

[127] J.T. Bushberg, J. Anthony Seibert, E.M. Leidholdt Jr., and J.M. Boone, *The essential physics of medical imaging*, Lippincott Williams & Wilkins, Philadelphia, USA 2002.

[128] I. Wendendelhag, Q. Liang, T. Gustavsson, and J. Wikstrand, "A new automated computerized analysing system simplifies reading and reduces the variability in ultrasound measurement of intima media thickness," *Stroke*, vol. 28, pp. 2195–2200, 1997.

[129] V. Zlokolica, Advanced nonlinear methods for video denoising, Ph.D. dissertation, Ghent University, Belgium, 2006.

Author Biography

Christos P. Loizou was born in Cyprus on October 23, 1962, and received his BSc degree in electrical engineering and Dipl-Ing (MSc) degree in computer science and telecommunications from the University of Kaiserslautern, Kaiserslautern, Germany and his PhD degree on ultrasound image processing of the carotid artery from the Department of Computer Science, Kingston University, UK, in 1986, 1990, and 2005, respectively. He is an assistant professor at the Department of Computer Science, School of Sciences, Intercollege, Limassol, Cyprus, since 2000. He is also a research scientist at the Department of Computational Intelligence of the Cyprus Institute of Neurology and Genetics since 2001. From 1996 to 2000, he was a lecturer in the Department of Computer Science, Higher Technical Institute, Nicosia, Cyprus. He has been involved in numerous projects in these areas funded by the Cyprus Institute Promotion and Foundation, such as the IASIS, TALOS, NEVRORAFI, OPTOPOIISI, and Visema, with a total funding managed close to 1 million euros. He has published 5 chapters in books, 5 refereed journal articles, and 20 conference proceeding papers in these areas. His current research interests include medical imaging, signal and image processing, pattern recognition, biosignal analysis, and computer applications in medicine. He was a conference session chair of the *IEEE Information Technology in Biomedicine*, ITAB06. Dr. Loizou is a member of the IEEE and a senior member of the Institution of Electrical Engineers. He lives in Limassol, Cyprus, with his wife and children, a boy and a girl.

Constantinos S. Pattichis (S'88–M'88–SM'99) was born in Cyprus on January 30, 1959 and received his diploma as technician engineer from the Higher Technical Institute in Cyprus in 1979, his BSc in electrical engineering from the University of New Brunswick, Canada, in 1983, his MSc in biomedical engineering from the University of Texas at Austin, USA, in 1984, his MSc in neurology from the University of Newcastle Upon Tyne, UK, in 1991, and his PhD in electronic engineering from the University of London, UK, in 1992. He is currently a professor at the Department of Computer Science of the University of Cyprus. His research interests include ehealth, medical imaging, biosignal analysis, and intelligent systems. He has been involved in numerous projects in these areas funded by the EU, the National Research Foundation of Cyprus, the INTERREG, and other bodies, such as the IntraMEDNET, InterMed, Future Health, Ambulance, Emergency,

ACSRS, Telegyn, HealthNet, IASIS, Ippokratis, and others, with a total funding managed close to 5 million euros. He has published 47 refereed journal articles, 110 conference papers, and 17 chapters in books in these areas. He is coeditor of the books *M-Health: Emerging Mobile Health Systems*, published in 2006 by Springer, and *Information Technology in Biomedicine*, to be published in 2009 by IEEE. He was guest coeditor of the special issues on *Emerging Health Telematics Applications in Europe* and the forthcoming *Emerging Technologies in Biomedicine* and *Computational Intelligence in Medical Systems* of the *IEEE Transactions on Information Technology in Biomedicine*. He was general cochairman of the *Medical and Biological Engineering and Computing Conference* (MEDICON'98) and the IEEE Region 8 *Mediterranean Conference on Information Technology and Electrotechnology* (MELECON'2000) and program cochair of the *IEEE Information Technology in Biomedicine*, ITAB06. In addition, he served as an associate editor of the *IEEE Transactions on Information Technology in Biomedicine* and *IEEE Transactions on Neural Networks*. He served as chairperson of the Cyprus Association of Medical Physics and Biomedical Engineering (1996–1998) and the IEEE Cyprus Section (1998–2000). He is a senior member of IEEE.